SEP 0.2 2020

D0151207

BIRD
LOVE

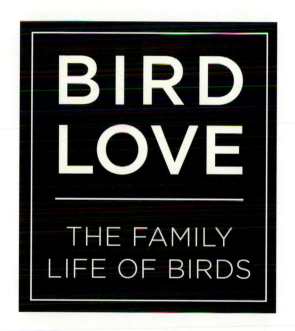

BIRD LOVE

THE FAMILY LIFE OF BIRDS

Wenfei Tong

Consultant Editor Mike Webster

Princeton University Press
Princeton and Oxford

First published in the United States and Canada in 2020 by
Princeton University Press

Princeton University Press
41 William Street
Princeton, New Jersey 08540
press.princeton.edu

First published in the UK in 2020 by
Ivy Press
An imprint of The Quarto Group
The Old Brewery, 6 Blundell Street
London N7 9BH, United Kingdom
T (0)20 7700 6700
www.QuartoKnows.com

Text copyright © 2020 Wenfei Tong
Design and layout copyright © 2020 Quarto Publishing plc
From an original idea by Wenfei Tong

All rights reserved. No part of this book may be
reproduced or transmitted in any form or by any means,
electronic or mechanical, including photocopying,
recording, or by any information storage-and-retrieval
system, without written permission from the copyright holder.

ISBN: 978-0-691-18884-3

This book was designed and produced by
Ivy Press
58 West Street, Brighton BN1 2RA, UK

PUBLISHER David Breuer
EDITORIAL DIRECTOR Tom Kitch
ART DIRECTOR James Lawrence
COMMISSIONING EDITOR Kate Shanahan
PROJECT EDITOR Joanna Bentley
DESIGNER Wayne Blades
PICTURE RESEARCHER Kate Duncan

Printed in Singapore

10 9 8 7 6 5 4 3 2 1

CONTENTS

FOREWORD

By Mike Webster

A pair of barn swallows has built a nest under the roof over my front porch. In fact, this same pair has nested on my porch for the past few years, returning each spring to build their cup-like mud nest in the exact same spot. This has given my wife and me an excellent front-row seat to the family life of these little birds. We see the male excitedly court and sing to his mate, wooing and gently coaxing her to start nesting. We watch them build up a nest from nothing, bit by bit with small bill-fulls of mud, then line it with feathers, eventually laying four brown eggs and then feeding the noisy, begging nestlings that emerge from those eggs. And after the young fledge and take to the wing themselves, albeit awkwardly at first, we watch the parents start all over again to squeeze in a second brood before the fall comes and they make the return journey to their southern wintering grounds.

Our swallows are the very picture of a cooperative, monogamous, and loving family. Yet, as a behavioral biologist, I know that lying beneath that cooperation is a lot of complexity, competition, and conflict. The male courts his mate with song and showy plumage, but he also courts and copulates with the mates of other males, and his own mate will likely copulate with "extra-pair" males herself. He will jealously guard her and repel rival males to keep that from happening, and possibly will reduce his own parental care if he suspects that some of the young in his nest are not his own. The young in the nest also compete with each other, each selfishly trying to extract as much parental care as possible. And the parents will conflict with those offspring on who gets the food, and also when to leave home.

That is all in just one species. Across the 10,000-plus species of bird that live on this planet, there is an amazing diversity of behaviors aimed at one simple goal: reproduction. This book is a collection of some amazing examples of these, from elaborate courtship displays that give potential mates information about genetic quality, to intricate nests designed to protect the young from predators and parasites.

We see loving parents, including same-sex partners that raise their young together, and individuals that forgo breeding altogether to help others instead. But we also see the "dark side" of reproduction: aggression, sneaky matings, deceit, infanticide, and parents that dupe others into raising their young. This book is about that behavioral diversity, and about the many ways that birds have evolved to pass their genes onto the next generation. I hope you enjoy reading it as much as I have.

INTRODUCTION

Bird family life can look rosy, from the long-term pair-bonds of parrots and albatrosses to the many species in which both sexes, and sometimes an extended family, share parental care duties. However, nature is amoral and a darker side, including sexual conflict, infanticide, and siblicide, is equally common. This book examines how and why both the light and dark sides of bird family life have evolved.

Bird reproductive strategies, from finding a mate to raising offspring, have evolved under a wide range of selective forces, producing diverse behaviors, from using cosmetics to kidnapping nannies. Birds are everywhere, and this book aims to deepen your appreciation of the birds you see every day, wherever you live, and also any new birds you encounter. It celebrates the

global diversity of avian reproductive strategies, from the almost universal infidelity among socially bonded birds to the benefits of long-term partnerships; from single-parent families to clans that rear the next generation cooperatively.

This book is peppered with analogies and the language of agency as shorthand for describing the long-term evolutionary benefits underlying bird behavior. Saying something like "birds will evolve to bring more food to chicks in response to louder hunger cries if the genetic variation for that behavioral rule is passed on to a larger proportion of the next generation" is rather a mouthful. To avoid giving you indigestion, I resort to giving birds agency. For instance, birds behaving almost like conscious investors of their genes in economic analogies are a recurring theme in this book.

The trouble with being a sexually reproducing organism is that one has to go to the effort of finding a partner with whom to invest in the next generation. An alternative, but not mutually exclusive, route is to perpetuate one's genes by helping relatives reproduce. Many organisms, including some humans, do this. However, regardless of whether the inherently cooperative venture of reproduction occurs in pairs or groups,

it is always subject to conflicts of interest because none of the members' genetic interests are completely aligned. For instance, a male bird may have less of a genetic stake in a brood of chicks than his mate, because she has mated with multiple males.

Although both parents have equal genetic shares in each chick, they seldom invest equally in resources that improve their chick's chances of surviving to reproduce. We will explore how the asymmetry in male and female investment in sex cells also explains many other sex differences.

Monogamy is the best reproductive strategy for Atlantic puffins.

Male and female Java sparrows look very similar, but only the male sings.

INVESTMENT ASYMMETRIES

Why are so many female birds a dowdy brown when their mates are covered in iridescent splendor, and why are single parents in the bird world usually female? The answers to both these questions come back to an early asymmetry in parental investment between females and males.

BELOW

As a result of asymmetries in parental investment, many male birds, like this Costa's hummingbird, are more flashy than their female counterparts.

Biologists define the sexes not by how they look or behave, but by the size of their sex cells. Females make larger sex cells such as eggs, whereas males make smaller sex cells such as sperm. In some organisms—including seed plants, mammals, or birds —this early asymmetry in parental investment is further exaggerated, so that females invest even more in nutrient-rich seeds, placentas, or egg yolks.

Choosy females and flashy males
The female reproductive strategy, with large-investment eggs, is quality over quantity, whereas the male strategy is typically the reverse. In the same way as, if you wanted to invest a large amount of cash in only a few blue-chip stocks, you would choose those stocks with care, females tend to be the choosier sex because they make a few large investments in their eggs, and it pays to carefully choose a high-quality mate.

The opportunity costs of a bad investment are much higher for a female's future reproduction than for a male's, for sperm are individually cheaper than eggs. That is why in most bird mating markets, females are the buyers and males must advertise through song and dance.

Who gets left caring for young?
The investment asymmetry from the time eggs and sperm are made often leads to larger differences later in birds' reproductive life. In many bird species, both sexes contribute to raising offspring, but imagine a female that has chosen poorly and ended up with a male that left to pursue other females the moment mating was over. She could give up the egg and start again with a different male, but because eggs are so expensive compared to sperm, this would cost her more in future reproductive output than it would a male in the same situation.

ABOVE

Most birds, including chickens, lay eggs that are 30–40% yolk—the part that provides nutrients to the embryo. For its body weight, a North Island brown kiwi has one of the largest eggs of any living bird, and 65% is yolk, making it an even larger investment than an egg of the same size that was mostly egg white.

In species where one parent is enough to rear offspring, whoever can leave first and safely will escape all the costs of childcare. In animals with internal fertilization like birds, males can always leave first. One reason males don't always leave is because many birds seem to require at least two adults to successfully raise chicks.

FAMILY LIFE EVOLVES

We all have a mother and a father, but not all of us will reproduce before we die. This difference in individual contributions to the future gene pool defines the winners and losers in an evolutionary battle to persist through our descendants.

ABOVE

There are five tody species, each endemic to a different Caribbean island. This Jamaican tody breeds in pairs, whereas the closely related Puerto Rican tody sometimes has non-breeding helpers at the nest.

For sexually reproducing organisms like birds and humans, the first step in producing genetic descendants is usually finding a mate. Optimal mating strategies will differ, depending on the sex, species, environmental conditions, and even age or condition of an individual. The first chapter of this book explores bird mating systems, from pure monogamy to pure promiscuity.

Chapter two discusses how birds attract and choose mates, and the remaining chapters elaborate on how they raise families. Chapter three concentrates on eggs and nests, whereas chapter four brings in chicks, which are independent, behaving entities in their own right. Chicks have their own long-term evolutionary interests, which do not always align with those of their siblings or parents.

The last three chapters focus on some of the more unusual ways birds have evolved to maximize their future genetic returns on family investment. Chapter five discusses situations where conventional sex roles are reversed, so that females are the larger, more competitive, and ardent sex, often with a harem of males who perform all childcare from incubation onward.

Chapter six looks at when care is extended beyond parents to other individuals, including many family members, but also unrelated adults with genetic shares in the communal brood. The final chapter discusses birds that rely solely on others to build nests, incubate eggs, and raise chicks. These brood parasites range from occasional parasites of their own species to highly evolved specialists that can only reproduce by duping a different species to care for their young.

Birds are the only dinosaurs still living—anyone who has
been chased by a turkey cock in full courtship mode will
be very aware of their striking resemblance to velociraptors
in *Jurassic Park*. Although flight is a relatively recent
evolutionary phenomenon, much of their reproductive
behavior, including nesting and parental care, evolved
much earlier among dinosaurs. Arrows in the chart below
indicate lineages still living today.

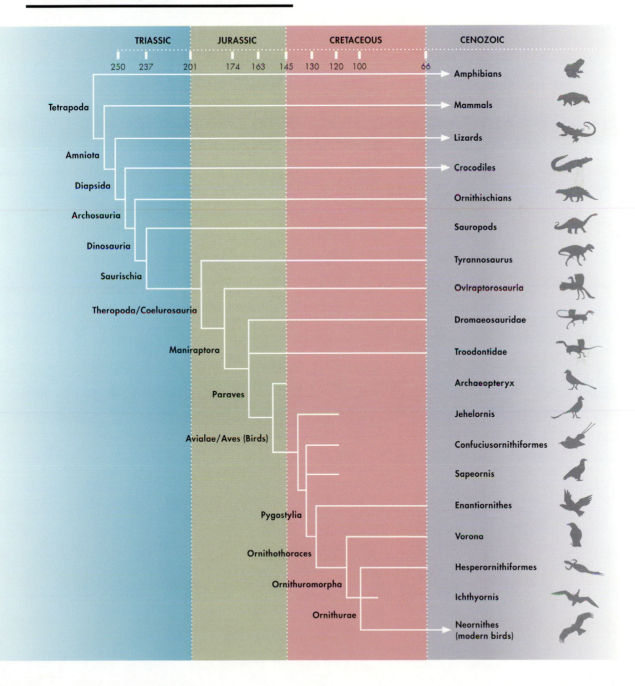

TRIASSIC	JURASSIC	CRETACEOUS	CENOZOIC
250 237 201	174 163 145	130 120 100 66	

Amphibians
Mammals
Lizards
Crocodiles
Ornithischians
Sauropods
Tyrannosaurus
Oviraptorosauria
Dromaeosauridae
Troodontidae
Archaeopteryx
Jehelornis
Confuciusornithiformes
Sapeornis
Enantiornithes
Vorona
Hesperornithiformes
Ichthyornis
Neornithes
(modern birds)

Tetrapoda
Amniota
Diapsida
Archosauria
Dinosauria
Saurischia
Theropoda/Coelurosauria
Maniraptora
Paraves
Avialae/Aves (Birds)
Pygostylia
Ornithothoraces
Ornithuromorpha
Ornithurae

Backyard Bird

To remind you that most of the diversity of bird family life in this book can be found right in your backyard, rather than in the wilds of Papua New Guinea or Antarctica, you will encounter a series of boxed examples like this running throughout the book. These boxes will feature examples involving common birds, some with an almost global distribution, such as barn swallows, and others that are more locally common, such as dunnocks and American robins.

Spectrums of behavior

Although humans tend to classify diversity in discrete categories, such as "male" and "female" sex roles, or "cooperation" and "conflict," much of this book is about variation in behavior across evolutionary time, within species, and during individual lifetimes. The paleognaths, which descend from birds that branched off early in the avian evolutionary tree, have predominantly paternal care, whereas for songbirds, biparental care is the norm. Brood parasitism may have evolved from communal breeders, and in greater anis some individuals specialize more in parasitizing other nests, while others specialize in cooperating. Individuals in a population vary in personality, which influences their mating and parenting behavior. In the case of white-throated sparrows, these behavioral differences are genetically inherited, but in others, like dunnocks, the same individual can flexibly alter its behavior from year to year.

Who wins?

Throughout this book, you will encounter examples of how bird family life involves both cooperation and conflict at all stages of life and levels of social organization. Some adaptations, such as nest structures perfectly contrived to befuddle predators, are the epitome of evolutionary efficiency. However, many aspects of social life, including reproduction, are fraught with counterproductive and maladaptive arms races. The unwieldy and disturbingly long genitals of ducks, infanticide, and siblicide are all examples of how internal conflicts within a cooperative group can lead to a suboptimal stalemate for everyone.

Ultimately, all these strategies and behaviors evolved because the genes that programmed them replicated more. You can imagine genes forming temporary alliances and cooperating to build complex organisms with behaviors that maximize the chances of those genes replicating. This gene-centered view of evolution implies that organisms are temporary and disposable robots, but robots are fascinating in their own right.

I hope a deeper appreciation for the diversity of bird behavior gives you a better understanding not just of birds, but also of the evolutionarily universal aspects of family life. Learning more about birds does not just teach us to view ourselves differently; it can also help us continue to coexist with as many of these diverse and fascinating beings as possible.

BELOW

A pair of great egrets building their nest in a Florida wetland.

ONE

ECOLOGY &
MATING SYSTEMS

What determines why some individuals or species make more faithful partners or who, if anyone, "wins" in a battle of the sexes? From blue-footed boobies (left) to backyard birds such as dunnocks and sparrows, this chapter explores the causes and consequences of bird mating systems.

MONOGAMY AND MORE

More than 90 percent of birds appear to be monogamous, in that males and females form pair-bonds and raise chicks together. However, what goes on beneath the surface of the "social mating system" is quite different. Males and females don't always want the same thing out of a reproductive relationship, and genetic mating systems are the outcomes of a battle of the sexes to leave the most descendants, played out in individual lifetimes and over generations of evolution.

MONOGAMY:
1 female + 1 male

POLYGYNY:
1 male + multiple females

POLYANDRY:
1 female + multiple males

POLYGYNANDRY:
Multiple mates for both sexes

Rules of engagement

Mating systems are defined by the number of partners each sex has. Monogamy for one female and one male, polyandry for one female mating with multiple males, polygyny for the reverse, and polygynandry for reciprocal promiscuity. This rather dry taxonomy belies a host of complications, but it is a convenient place to start exploring and explaining the diversity of bird mating systems.

ABOVE

Scientists classify mating systems into four main categories. This chapter explores why some species or individuals tend to have a particular mating system.

RIGHT

The forest-dwelling, insectivorous blue-billed malimbe is a monogamous weaver species.

FAR RIGHT

Other weaverbirds, such as the village weaver, are polygynous, keeping several nests for several different mates.

At the broadest levels of biological organization, major groups of birds tend to be more monogamous if their chicks require more care. A young eagle or albatross can't survive without the care of two parents, who are part of a stable, long-term relationship. In contrast, ducklings are so independent they can feed themselves from the moment of hatching, and we see little parental investment by most male ducks.

Closely related birds from the same genus have different mating systems largely because of what they eat and where they live. For instance, forest-dwelling weaverbirds are generally insectivorous, and remain in monogamous pairs that guard territories all year round. It takes two adults to catch enough insects to feed a hungry brood.

Polygyny is most common in fruit and seed eaters because the seasonal abundance of these foods makes them hard to monopolize, and allows just one parent to provide enough to raise chicks. Males can usually afford to desert first, because as long as sex ratios are even, males stand to gain more offspring by remating. Females are typically more constrained by the number of offspring they can produce, so they gain less by deserting their current family for another mating. In these polygynous systems, some males manage to win many matings while others don't breed at all.

LEFT

Most birds of prey, like these bald eagles, need both parents to successfully rear offspring, and form long-term pair-bonds.

Within species, mating systems are the passive outcomes of social conflicts between individuals attempting to maximize their own reproduction. Natural selection favors those who pass on the most genes, but there is an inevitable trade-off between how many offspring one can produce, and how many one can rear to maturity. Investing in the former, and acquiring more matings, usually takes place at the expense of the latter. The resulting mating system still depends on ecology in two ways. Firstly, the distribution of food influences how potential mates are distributed over space and time, which in turn alters how easily one can find and keep a mate. Secondly, food and habitat also alter the costs and benefits of investing in parental care, which determine whether one parent can desert to seek additional mates.

BELOW

The availability of food in a habitat such as savannah dictates the mating system. Grey-capped social-weavers nest in colonies.

WINNER TAKES ALL

Competitive groups

Let's start with leks. Leks are arenas where males gather to display to females. After a very brief sexual encounter, females complete reproduction solo, from nesting all the way to raising chicks. This system, found in a handful of birds including grouse and birds-of-paradise, is an extreme version of polygyny. One or two males win almost all the matings, while the vast majority contribute nothing to the next generation.

Males are unable to monopolize females when food is spread out, so it pays more to gather in one place and advertise their attractiveness to potential mates, forming a lek. As females don't get anything from males other than good genes, an opportunity to view a gallery of potential sperm donors simultaneously makes it easier to choose the best genetic option. The courtship dynamics are not unlike those at a pop concert. From the viewpoint of males, even lowly males stand a better chance of mating by hanging on the fringes of a lek than by going solo.

Monopolizing resources

Monopolizing a limited food source can be an effective method to attract multiple females. For yellow-rumped honeyguides, beeswax is a much sought-after delicacy, and bee hives in the Himalayas are in short supply and sparsely distributed, so one male can control access to a single bee hive. Females flock to these top territories for the wax, and copulate with the territory owners before going off to raise the chicks on their own. In this species, a single male has been observed copulating 46 times with at least 18 different females. As with leks, the top males dominate the mating market, while lowly males without a bee hive have little chance of breeding.

TOP

Sharp-tailed grouse, like many members of the chicken order, form leks in which males display with dances and colorful ornaments to attract mates.

ABOVE

Yellow-rumped honeyguides have a polygynous mating system based on resource defense, whereby males attract multiple mates by guarding sought-after bee hives.

21

Two parents or one

When two parents are needed to successfully raise a brood, neither party can be left rearing the young solo, but resource distributions still dictate mating systems. When resources are clumped, like the reeds that red-winged blackbirds in North America nest among, a male's best bet isn't to head off to the concert arena of the lek, but instead to establish an attractive property. Red-winged blackbird females prefer to be the second, or even third, mate to a male with an outstanding territory than to monopolize the parental help and attention of a male with less material wealth to offer.

In contrast, yellow-headed blackbird females feed in fields away from the marshes where they nest, and males offer almost no parental help, so there is little cost to females in nesting close together and sharing a mate. When male care is minimal, where females settle is purely a function of the habitat, and a mating system emerges based on how clumped and defendable females are.

Out of 122 well-studied European songbirds, 40 percent show occasional polygyny. Male desertion in these species is possible because, although a female raises fewer chicks than if she had his help, she can still raise about half her chicks alone. If a male can have more offspring by mating with more than one female to compensate for their lower survival due to his lack of care, it pays him to have two families. In many of these species, including warblers, flycatchers, and sparrows, biologists have induced polygyny by removing some males and tempting their remaining neighbors to set up house with more than one female, whom they either abandon to return to their first mate, or help part-time.

LEFT

Male yellow-headed blackbirds owe their bold plumage to an evolutionary investment in attracting multiple mates, rather than in parental care.

TOP

Female yellow-headed blackbirds are solo parents, and feed far from their nests, minimizing competition for paternal care and nest sites.

SHADES OF MONOGAMY

Unlike the majority of mammals, 90 percent of bird species are socially monogamous. These birds behave rather like human couples. They may go through a series of formalized rituals, akin to an exchange of marriage vows, or they may simply establish a property together and share the work of childcare. To the outside observer, they constitute a couple.

Things get more complicated, however, because social contracts can be very short lived, or can be broken—either openly, or through cheating. Individuals can change their strategies from one breeding attempt to another, switching flexibly from monogamy to polyandry, polygyny, or even polygynandry. In some songbirds, this degree of flexibility is the norm. For others, like many large seabirds or raptors, the social bond typically lasts a lifetime, and "divorce" is rare. In between, there is serial monogamy, where couples stay together for one year or more, but then find other reproductive partners.

BELOW

Infidelity exists, even among large seabirds like these black-browed albatross that typically form socially monogamous pair-bonds for decades.

Most bird species form monogamous pair-bonds for a single season, but biologists refer to this as social monogamy because cheating is rife. Biologists only noticed this with the advent of the same DNA fingerprinting technology that humans use in paternity tests. Of all the bird species studied, three-quarters have what biologists call "extra-pair paternity," or EPP. In other words, females in socially monogamous couples frequently philander with males other than the one who is helping them to defend a territory and raise chicks. The biological literature is now littered with EPPs, distinct from EPCs (extra-pair copulations), terms that have proliferated since the discovery that cuckoldry is so common.

Backyard Bird

DUNNOCKS

Drab little dunnocks in the suburbs have some of the most exciting mating system dramas among birds. This is because both sexes rear the most chicks by mating with as many partners as possible, while also retaining exclusive access to their individual mates. Dunnock females first establish exclusive territories, and males do their best to adjust. Females must defend larger territories when food is more spread out. If two males are unable to monopolize a large female territory, their territories coalesce over time, so that by the time she is ready to copulate, the two males share and jointly defend a territory that overlaps the female's. In this situation, the alpha male attempts to guard the female to prevent her from mating with the junior male. In contrast, the female does her best to elude this mate guarding so she can mate with both males. If she fails, the beta male could destroy her eggs and chicks, in which he has no genetic share. If she succeeds, she raises more chicks by ensuring paternal care from both males.

WHEN TO CHEAT

The occurrence of multiple mating varies across species and within populations. Larger, long-lived species, which have few offspring and take a long time to mature, show lower levels of EPP than species with faster life histories. In addition, species where a male's care is less necessary tend to show higher levels of EPP.

Perhaps this is because cuckolded males can only retaliate by caring less for chicks if females need their help. In species where females can do perfectly well alone, it pays more to cheat and gain better genes for the offspring than remain faithful to gain extra childcare from one's genetically inferior mate. In serins, small Eurasian songbirds, experimentally liberating both parents by providing extra food resulted in males caring less for chicks and increased EPP rates.

Within species, individual birds use cues like population density and the availability of better options to flexibly alter the degree to which they cheat on their social mate. This is because most species, from blue tits to chickadees, go next door for EPCs, so opportunities to cheat depend on how accessible and available the neighbors are. Eastern bluebirds with lots of nest boxes densely packed together have higher levels of EPP than bluebirds that are more spaced out.

BELOW

Living in close proximity with other Eastern bluebirds makes extra pair copulations more likely.

None of this explains why females cheat, especially when doing so is costly because of punishment by a cuckolded mate, either in the form of reduced childcare or aggressive mate guarding that interrupts her attempts to feed. Yet biologists have observed female blue tits, hooded warblers, black-capped chickadees, and red-winged blackbirds making forays off their territories to consort with neighboring males. There is some evidence that females are most tempted by a quick fling with a much more attractive specimen who could be older, larger, more ornamented, have a higher social standing, sing better, woo her with prettier flowers, or any combination of the above.

The reproductive results of these flings explain why females have evolved a tendency to cheat. Blue tit chicks sired by extra-pair males survive better than their half siblings, and extra-pair bluethroat chicks have stronger immune systems. In savannah sparrows and reed buntings, chicks sired by extra-pair males are healthier, though there is no evidence that EPP benefits female Darwin's finches, sand martins, or coal tits.

BELOW

Blue tit chicks fathered by extra-pair males survive better than their half siblings. This genetic advantage could explain why females engage in extra-pair copulations.

AUSTRALIAN SUPERB FAIRYWRENS

Australian superb fairywrens would win the world record for cuckoldry. Ninety-five percent of broods have extra-pair chicks. During the day, females stay on their territories and receive the attentions of neighboring males, who woo extra-pair females (but not their own mates) with flower petals that tend to complement their blue plumage. There is no relationship between how many times a male courts a female and the number of chicks he sires with her, suggesting that females are in complete control of adulterous liaisons. Exactly when extra-pair copulations occur had been a mystery, because copulations are rarely observed despite frequent extra-pair courtship. It was only by radio-tracking females that biologists discovered the answer. Just before dawn, reproductive females often sneak over to the neighbors they found most attractive, and are then back home by daybreak.

Shorebirds provide a spectrum of mating systems that are mostly determined by how much care chicks require, and which sex is more likely to make it to adulthood. As young shorebirds go, oystercatcher chicks are rather helpless, requiring their parents to remain in a monogamous pair-bond for long enough to raise the offspring together. In contrast, other species, like the Kentish plover (formerly the same species as the endangered snowy plover in North America), have highly precocial chicks that are able to swim and run for three quarters of a mile (just over 1 km) within hours of hatching.

Kentish plover chicks only need one adult to raise them, which means that their parents are caught in a tug-of-war over who gets left caring for the young. The parent that can abandon the family first "wins" in an evolutionary sense if they can increase their mating opportunities and produce more offspring. The parent that was abandoned has to stay and compensate for the loss of a partner by caring for their joint reproductive investments.

In Kentish plovers, females desert more often than males. The reason is not because single males do a better job than single females, because experimentally removing either sex causes no difference in the number of offspring fledged. Instead, there is a big difference in how long it takes the sexes to find a new partner. Females that were experimentally removed from their families and "forced" to try again remated within two days, whereas males took an average of 12 days to find a new mate. This is a numbers game, in which the rarer sex is always in higher demand. An equal number of male

and female chicks hatch out of eggs, but, for reasons that are not entirely clear, more sons make it into adulthood, resulting in an excess of males, all of whom find it relatively hard to find a new mate. As a result, male Kentish plovers are currently the ones that do most of the childcare, while females are free to move from male to male.

Across shorebird species where one parent is all that is needed, polyandry is most often associated with high breeding densities and very long migratory routes. For instance, white-rumped sandpipers, in which females usually leave males to take care of the chicks, breed beyond the Arctic Circle in Alaska and Canada, and migrate all the way to Patagonia for the winter. Perhaps laying eggs is such an expensive business that females are unlikely to have enough energy left to migrate if they participate in childcare as well, so they simply leave earlier, thereby "forcing" males to pick up the slack.

ABOVE

Eurasian oystercatcher chicks need the care of both parents, so neither sex can afford to desert the family until their chicks have fledged.

LEFT

Kentish plover chicks are independent enough to survive with the care of a single parent. The rarer sex is more likely to remate, so tends to desert first. This is usually the female, which means that male Kentish plovers are usually left taking care of the chicks on their own.

POWER AND STATUS

Ravens, like humans, engage in a lot of intricate political maneuvering throughout their lives. Raven couples keep track of the ever-changing relationships between other individuals across years, often using this knowledge to manipulate the behavior and status of competitors.

A long-term mated pair ranks in the top echelons of raven society, whereas young floaters without friends are at the very bottom, and those in the early stages of pair-bonding (with either sex) are intermediate in the hierarchy. What is most striking is that power couples selectively disrupt budding relationships, probably as a way to maintain status. The longer a couple has been together, the more time they spend sabotaging the potential relationships of others.

LEFT

Raven power couples are long-term monogamous pairs with a reproductive advantage because they dominate less pair-bonded ravens at shared resources such as carcasses.

Backyard Bird

AUSTRALIAN BLACK SWANS

Australian black swan couples with the highest social status have the best territories with the most food, and raise the most chicks. They use their curly black feathers as badges of status, and couples show off these feathers in "triumph displays." Individuals tend to choose long-term mates based on how many curly feathers both partners have, so the most attractive females mate with the most attractive males, and command the most status in black swan society.

BETTER THE DEVIL YOU KNOW

Like ravens, albatrosses can take a long time to find their life partner. In some species, the average age of first reproduction is 10 years, and some individuals don't start breeding until they are 20 years old. Albatrosses can live for more than 40 years, so can afford to take one to four years to establish a strong pair-bond before laying an egg.

Albatross reproduction is also highly expensive for both sexes. Females of many large species can afford to lay just one egg every two years, and two adults are needed to support this single offspring. You would be forgiven for assuming that albatrosses are the epitome of long-term monogamy, and they are, but not of fidelity. Albatross divorce rates are almost nonexistent, but wandering albatrosses, one of the largest seabirds, have up to 25 percent of chicks sired by a male other than the one that is rearing them. Sexual infidelity is also present in the waved albatross of the Galapagos, and other smaller albatross species that have been subjected to paternity tests. Similarly, sandhill crane and black swan couples experience infidelity but seldom divorce.

However, males don't really have the option to desert in these large, long-lived birds because two adults are needed to raise any offspring. Furthermore, long-term couples tend to raise more chicks. In addition to having higher-quality territories, more faithful couples can start nesting earlier, because they take less time to get through all the ritualistic endearments and tokens of commitment before getting down to the business of breeding. For migratory species with a short window of opportunity to breed, starting early can make all the difference to whether your chicks have time to reach independence before the weather turns ugly. Long-term couples are also more efficient at feeding their chicks than newlyweds.

BELOW

Wandering albatross are one of the two largest albatross species, and engage in elaborate pair-bonding rituals every two years, when a long-term couple reunites to breed.

BELOW

Zebra finch couples that have been together for longer raise more chicks than newlyweds, because they start breeding earlier in the season.

Often, divorce is simply not worth the time and effort it takes to find and establish an efficient parenting relationship with a new partner. Long-term Australian gannet couples raise more chicks than those that keep divorcing and re-pairing. Pintail ducks that remain bonded over the winter save time and effort on courtship. Black-legged kittiwakes that have to find a new partner have higher stress levels and raise fewer chicks than those that stick with a long-term mate, even if some of their chicks are the products of infidelity. Tiny, fast-breeding songbirds like zebra finches are more conducive to laboratory experiments than albatrosses, so we know that established couples get down to business sooner and raise more chicks than newlyweds. Zebra finches breed in the Australian desert, giving them a narrow window of opportunity after the rains in which to raise chicks. Just like in penguins, which must raise chicks before the ice melts, long-term monogamy is the best strategy for zebra finches because familiarity saves precious time.

Blue-footed boobies live for more than 20 years, engage in reciprocal courtship which involves both parties showing off their bright blue feet, and both parents contribute to raising chicks. However, even in these long-term couples, females are not always faithful, and sexual infidelity peaks in the second year of breeding together. Perhaps this then tapers off again because males subsequently cotton on to their mate's infidelity and retaliate by attempting to guard her. By destroying suspect eggs, males can make adultery less worthwhile for their mates as the pair become more intimately acquainted over the years.

LEFT

Blue-footed boobies form long-term pair bonds, renewed every year by a ritual display of their blue feet.

BILLING AND COOING

Birds that establish and maintain long-term pair-bonds use a variety of gestures and sounds to reinforce the strength of the relationship. This helps them to function as an efficient reproductive team, and to achieve their mutual evolutionary goal of leaving as many descendants as possible.

Parrots maintain long-term monogamous pairs despite adulterous undercurrents, and vocal communication plays a large part. For example, female budgerigars remember their mate's voice for months after the couple are experimentally separated. They only begin to display an interest in the calls of new males when the memory of their mate's voice begins to fade.

Mated pairs also use their voices in other, slightly more practical ways. Although male birds do the majority of territorial singing in temperate zones, female birds in the tropics often sing just as much as males. Duets can function as coordinated territorial displays to ward off intruders, or as contact calls, for instance in rufous and white wren pairs that have to find each other in the dense rainforest understory.

Bonded birds also engage in a great deal of allopreening—using one's bill to groom and caress another individual. There is evidence from over 500 species that allopreening between mated pairs is more common in species where both parents raise the offspring, and couples that preen each other more have a higher chance of remaining together over the years. Rates of allopreening have no bearing on the rate of sexual infidelity, only on divorce rates between socially monogamous couples.

TOP

Many seabirds, such as black-browed albatross, live in large colonies. Yet they can recognize their mate's voice from the cacophony of other calling couples.

ABOVE

Allopreening—when one bird preens another—helps to cement pair-bonds, such as between this pair of rockhopper penguins.

DIVORCE

Almost all socially monogamous bird species that establish pair-bonds, even for a breeding season, experience divorce. Regardless of whether the decision to divorce is mutual or initiated by one partner, there is evidence that, at least for one of the parties, divorce improves reproductive success.

In the 8 percent of oystercatcher pairs that divorced naturally, individuals who initiated the divorce usually moved to a better territory and raised more chicks, whereas the victims of the divorce raised fewer chicks than in their original pairing. The jury is still out on whether the propensity to divorce is genetically influenced among species that are largely monogamous.

A more direct test of reproductive success triggering divorce took place among captive canaries, where females were presented with a choice between their former mate and a former neighbor, whom they knew but had never reproduced with. The former mate reacted to the sight of the female by gathering nesting material, while former neighbors sang lustily. Females only showed a preference for the neighbor if they had failed to raise more than one chick with their former mate.

Divorce rates can also vary across populations of a single species. In blue tits, some populations have divorce rates as high as 85 percent, whereas in others only 5 percent of pairs pair up with new individuals. One of the biggest predictors of divorce in this species is when pairs return to the breeding grounds. Pairs that arrive at very different times are more likely to divorce, which makes sense, particularly when the breeding season is short and there isn't time to hang about waiting for one's partner from last year.

RIGHT

Divorce in herring gulls is often triggered by a failed nesting attempt.

BATTLE OF THE SEXES

Although sexual reproduction is essentially a cooperative investment in the next generation, like most partnerships in which the long-term strategies of both parties differ, there is room for conflict and cheating. At times, this conflict can escalate into an arms race between the sexes.

While most birds have never evolved anything more cumbersome than a simple genital opening called a cloaca, ratites (ostriches, emus, and their kin) have retained something akin to the phalluses on early birdlike dinosaurs. However, in a minority of birds, such as ducks, the phallus has evolved to spectacular and mind-boggling lengths, and the best explanation for this ungainly structure is sexual conflict.

Male ducks later in the breeding season are at something of a loose end, and eager to gain a couple of extra offspring at comparatively low cost. In contrast, females have already reproduced with the mate they chose earlier in the season, and have no evolutionary interest in being forced to copulate with a random cad who isn't even going to help with the childcare. The outcome of this male-biased sex ratio is forced copulation, which is sometimes fatal to the females, who can drown.

Another outcome is the evolution of extraordinarily long genitalia. Consider early competition among male ducks, when even the tiniest nub of an intromittent organ could give a male an edge over the others with cloacas. Over time, one can imagine an escalation in which the ducks with the longest phalluses produced the most offspring. However, this doesn't quite explain the full story, including why duck phalluses are sometimes not straight, but corkscrewed.

It took a female biologist to ask what on earth was happening in the female reproductive tract. This scientist found a farm that raised hybrid ducks for *foie gras*, where there were many drakes accustomed to ejaculating into test tubes for artificial insemination. She gave them three kinds of test tubes to ejaculate into: straight, corkscrewed in the same direction as

their phallus, and corkscrewed in the opposite direction. Phalluses get stuck in the first turn when attempting to penetrate a test tube corkscrewed in the opposite direction, and that is exactly what female ducks have evolved to counteract unwanted inseminations.

The rather baroque genitalia in male ducks is matched in both length and shape by female genitalia in an evolutionary arms race between the sexes, where longer phalluses have driven the evolution of longer and more convoluted vaginas. Comparing across duck species, the more frequently forced copulation occurs, the longer the male phallus, and the more pouches (dead ends for sperm) there are in the female's reproductive tract. Like so many examples of elaborate and apparently maladaptive structures, duck genitalia are the evolutionary result of an escalating arms race, at the end of which both sides would have been better off if the race hadn't begun.

BELOW

The record holder for phallus size among ducks is the male Argentine duck, which has a phallus as long as his body.

TWO

COURTSHIP

Manakins dance, peacocks flaunt long plumes, and birds-of-paradise do both, whereas in other species, such as barn swallows, the signals of sexual attraction are more subtle. Why do some species evolve significant weapons or ornaments, while others do not? And, if you are on the receiving end, why prefer some courtship displays over others?

EXTREME SEX DIFFERENCES

Why is it that in many animal societies, males compete with one another for the favors of females, which are usually the choosy sex? Is this simply the product of human stereotypes imposed on the thinking of biologists, or is there a sound explanation for the evolution of divergent sex roles that involve male combat and female choice?

The theory of sexual selection was proposed by Charles Darwin, first in *Origin of Species*, and then more fully in *The Descent of Man*. With sexual selection, Darwin sought to explain why males and females look so different in many species, and why some of these sex differences involve structures or behaviors so elaborate that natural selection for survival could not possibly explain their existence. Biologists today still follow Darwin's categorization of sexual selection into competition within one sex (typically males) for access to the other sex, and selection by members of one sex (typically females) for attractive displays by the other. These are often referred to as male–male competition and female choice, and together they explain a large number of differences between the sexes in appearance and behavior.

The asymmetry in reproductive investment strategies explains why females, with fewer, higher quality investments are typically the choosy sex. Males compete for the relatively scarce reproductive investments of females. However, because we all have one father and one mother, if a few most competitive males win most of the females, the majority of males fail to breed. A situation with a few winners and many losers explains why sexual selection is usually stronger on males than females. It also explains why males in species with stronger sexual selection are less inclined to care for the offspring. Parental care would take time and energy away from a male's future success as a competitor for mates.

TOP LEFT

Male members of the chicken and pheasant order, such as grouse, turkeys, or this Temminck's Tragopan, often use colorful, inflatable skin to attract females.

LEFT

In shorebirds such as woodcock, snipe, or common greenshank, males engage in displays of agility to win females, taking steep aerial dives and zigzags.

TOP PERFORMERS

In a minority of bird species where females are single parents, males often gather on leks, where some of the most intense forms of sexual selection are played out. The reason lekking males like birds-of-paradise look and behave so differently from females is that there is an extreme asymmetry in parental investment, and a correspondingly high skew in male mating success. Only the sexiest males—those with the most extreme courtship displays— attract the majority of females.

Why do any species lek at all, given the potential costs of leaping about noisily in a large gathering? One would expect the predators to come flocking, but in lekking species, the benefits of group courtship must outweigh the costs. Biologists have come up with many hypotheses to explain the existence of leks, which we discuss overleaf. None of these are mutually exclusive, although they each focus on different beneficiaries.

BELOW

The King bird-of-paradise has an elaborate courtship dance in which he flaunts elongated tail feathers that resemble wires and fluffs up his white abdominal feathers until he looks like a cottonball bobbing in the forest.

Reasons for leks

Leks could form in response to predation, simply because each individual's chances of being eaten are lowered in a crowd, even if the crowd attracts more predators than an individual. This can even result in mixed-species leks. Greater prairie chickens (picture 1 below) tend to join the leks of the relatively sharp-eyed sharp-tailed grouse, to piggyback on the greater vigilance of the other species.

The hotspot hypothesis looks at things from the viewpoint of all males. If males, such as long-tailed manakins (2), cluster in places females frequent, they have a higher chance of being seen by a potential mate.

In contrast, the hotshot hypothesis benefits inferior males most, because it assumes that if they displayed on their own, they would attract no one at all, but if they hang about with the alphas, they might benefit from all the reflected glory and attract the odd female. When biologists placed fake model hotshot males in the middle of a little bustard lek (3), both sexes were attracted to the idealized models.

The female preference hypothesis assumes females can make the best choices when all the options are gathered together for easy comparison. This usually benefits the top males most, because females find it easiest to pick out the male with the best courtship display when competitors surround him. Even in lance-tailed manakins (4), which display within earshot but out of sight of neighboring males, all males got more female visitors by being closer to their neighbors, but only the top males got more matings as a result.

In many members of the chicken family, such as turkeys (5), males can benefit from lekking with relatives, so that even if they don't get to mate with any females, their genes are still more likely to be passed on through their brothers than if they displayed alone. Mysteriously, peacocks can recognize and lek with brothers they have never before met.

LEFT

Whatever the likely reason for lekking, what all leks have in common is a place where individuals gather to display to prospective mates.

FEMALE CHOICE

Why do females prefer certain ornaments over others? There are three main explanations. The first is that males have evolved to exploit preexisting female preferences for something like a bright color, and are psychologically manipulating females into wanting to mate by showing off something they already find irresistible for other reasons.

The second explanation is that the extravagant, often cumbersome ornaments are honest signals of quality because only a really healthy male with excellent genes can afford such an expensive handicap.

The third is often referred to as the "sexy son hypothesis." This explanation proposes that the only reason an exaggerated ornament evolves is because females that choose to mate with the flashiest males will have the sexiest sons, and therefore, the most grandchildren. The beauty of this idea is that rather than invoking some inherent genetic excellence, it relies purely on arbitrary preference and fashion, even at a cost to survival. There is some evidence that arbitrary female preference for a certain "look," without any additional genetic benefits, is enough to explain the evolution of ornaments such as the exaggerated tail on a peacock. These hypotheses are not mutually exclusive. Long tails could be honest indicators of quality and also make sons "sexy" if females have a preexisting predilection for them.

LEFT

The long tail of the male golden pheasant is a feature that can attract females.

HONEST SIGNALS

Like flashy sports cars, the displays that male birds employ to attract females are often costly to acquire and maintain, resulting in a trade-off between investing in mate attraction or individual maintenance. As a result, extravagances can be an honest signal of quality, and a reliable rule of thumb for birds to find the best genes for their offspring is to choose mates with the most exaggerated courtship displays.

BELOW

The courtship display of a male long-tailed widowbird in the East African savannah can be seen from half a mile (nearly 1 km) away as the male flaps slowly above the grass, showing off his 16-inch-long (45 cm) tail. Females, who are brown with short tails, prefer to mate with males with the longest tails.

Tails are often a good indicator to use in choosing a mate, partly because they are expensive to acquire and maintain. Bearded reedling females are less faithful if they have a neighbor with a longer tail or beard than their social mate. Black-billed magpies of both sexes prefer mates with longer tails, and magpies with longer tails pair and breed earlier, resulting in more offspring. However, only older birds or males with stronger immune systems invest in longer, heavier tails, because it's harder to keep a larger tail immaculate. Similarly, female barn swallows prefer males with symmetrical tails, because stress in early development is more likely to lead to asymmetric tails, which act as an indicator of weakness.

LEFT

Peacocks attract more females if they have more eyespot feathers on their iconic trains, and the chicks sired by peacocks with more eyespots survive for longer.

ABOVE LEFT AND ABOVE

Redder scarlet rosefinch (left) or red-backed fairywren (right) males have more extra-pair offspring because they attract more extra-pair copulations.

In blue tits and collared flycatchers, the males with the best ornaments survive better and attract the most extra-pair females. Their offspring also survive better than the chicks sired by each female's social mate. By mating on the sly with the sexiest males, females often seem to be shopping successfully for higher-quality genes. Conversely, females with more attractive social mates (as measured by larger ornaments or song repertoires) tend to be more faithful. A study across 73 species found that the higher the rate of extra-pair paternity, the more bright and showy the males were compared to females. However, the number of social mates made no difference, suggesting that males have evolved colorful plumage largely to woo extra-pair females.

The Red Queen effect

One of the main explanations for why honest and showy signals are sexually selected is because birds and other sexually reproducing organisms are in an evolutionary arms race with germs. The idea is named after the Red Queen in *Alice Through the Looking Glass*, who says, "it takes all the running you can do, to keep in the same place. If you want to get somewhere else, you must run at least twice as fast as that!" Birds must constantly evolve defenses against parasites, which in turn causes parasites to evolve new ways of invading their host's immune system. A general indicator of health, such as a long tail or bright red ornament, enables birds to choose the healthiest mates with the best genes.

Red badges of quality

Carotenoids—the pigments in carrots—are antioxidants that are important for maintaining good health and can make bird skin or feathers redder. However, birds can only get carotenoids from their food, making these pigments limited, and posing a potential trade-off between investing in attractiveness and health. You can only afford to squander carotenoids on looking good if your system is strong enough to handle the cost. For instance, redder house finches are more attractive to females and have a stronger immune system and fewer parasites. Feeding zebra finches fewer carotenoids makes them less red, less attractive, and less healthy all at once.

In monogamous greater flamingos, it is the females that use carotenoids as honest signals of quality. Redder flamingos are better mothers and more attractive to males. Female flamingos can flexibly change how they use precious carotenoids. They deposit a lot of carotenoids into feathers while growing them, well in advance of courtship, but save some for a secretion that comes out of a preen gland in their rear, then use their cheeks like brushes to apply the "rouge" to their neck and body feathers. This enables a female flamingo to flexibly enhance her already pink feathers.

ABOVE

Whiskered auklet facial plumes are sexually selected ornaments, but they also function as whiskers to help the birds navigate in burrows.

LEFT

Black guillemot couples tend to pair off by foot redness, which is an honest indicator of health.

RIGHT

Raptors like this osprey are usually socially monogamous, and pairs copulate at much higher rates compared to most birds—possibly a strategy by both partners to keep their mate too busy to be unfaithful.

Food for sex

Commitment is an unenforceable contract, but birds can signal their ability and willingness to provide by bringing gifts of food during courtship. Ospreys continue copulating for about 45 days before the female lays her eggs. During this time, males present their mates with presents of fish, and are rewarded with sex. Females judge both the frequency and quality of these courtship gifts, and males that bring smaller fish or feed their females less often don't even attempt to solicit sex as much, and are seldom rewarded. Biologists don't know how conscious birds are of these decisions. The more time a male has to devote to chasing off other males, the less time he has for fishing, so the frequency of gifts is a good measure of what a good defender the male is, in addition to his abilities as a provider. There is no evidence of cheating or false advertising, in which males that brought wonderful courtship presents turned out to be duds at providing once the offspring came along.

Courtship feeding is also common in terns, and usually peaks just before egg laying, when the female is heaviest and can't fly very well. Males bring food and the quality of eggs is higher if the food is better and more plentiful. Female whiskered terns take this a step further and use the promise of sex to lure other males into giving up the fish intended for their own mates. The females almost never actually follow through with copulation; they simply present their cloacas to a passing male to solicit his interest and his gift of fish.

BRAINS BEFORE BEAUTY AND BRAWN

In contrast to their elaborately ornamented relatives, the birds-of-paradise, male bowerbirds rely on artfully constructed galleries to attract females. The bowers they construct are not nests, for female bowerbirds are single parents, just like female birds-of-paradise. There are multiple lines of evidence that sexual selection has driven the evolution of brains rather than bodies as sexual attractants in male bowerbirds.

ABOVE

Satin bowerbirds have an innate attraction to blue objects and an aversion to red objects.

RIGHT

Bowerbirds of different species construct bowers of different levels of complexity, from a simple clearing to elaborate court structures.

Bowerbirds of different species build display courts of varying complexity, with the simplest differing little from the cleared courts in which birds-of-paradise dance. But the more elaborate bowers can be so large and complicated that there are tales of people in Papua New Guinea mistaking them for shrines built by a neighboring tribe. These complex bowers can be long avenues of woven twigs, or huts (some large enough to fit a small child) constructed around a central supporting maypole.

Around these structures, males display an assortment of treasures in attractive piles, and can be quite fussy about the color and arrangement of their collectibles. Bower construction requires a great deal of learning and practice, such that younger males will spend their first few years apprenticing with an older male. In one case,

TOOTH-BILLED BOWERBIRD

Cleared ground

ARCHBOLD'S BOWERBIRD

Thick mat of ferns

SATIN BOWERBIRD

Simple avenue with display area at one entrance

LEFT

Great bowerbirds have long avenue-styled bowers with a court lined with gray and white decorations such as pebbles, bones, and shells. Males lure the female into the bower and she looks out toward the court where he displays. By placing larger pebbles farther away from where the female sits, males frame the scene with an optical trick that flattens what a female sees. When experimenters reverse this gradient, males will restore it within two weeks.

a young spotted bowerbird learned to collect blue objects from his satin bowerbird mentor, rather than the red, green, and white typically preferred by his own species.

The rarity of a color tends to make it especially attractive, so male bowerbirds are often competing for a limited and precious resource when they search for items to add to their galleries. As a result, it is not uncommon for males to pilfer from or destroy one another's bowers. In addition to keeping track of who lives where and who has what, males are almost obsessively compulsive in their ability to remember where they placed each item in their galleries. What may look to the human eye like a pile of blue bottle caps is a lovingly

arranged display of female attractants for a male satin bowerbird. He will fastidiously rearrange anything that biologists have tampered with.

But females don't just want a mate dripping with machismo and talent; they expect him to be sensitive to their feelings as well. Female bowerbirds are quite skittish, and startle easily. In general, they prefer males with ardent and intense displays, but only if the male is quick to back off if she tells him that he is coming on too strong. Otherwise, the female flies off before he has had a chance to mate. If a male is able to adjust his courtship display quickly in response to a skittish female, he has a much higher chance of persuading her to stick around long enough to mate with him.

MACGREGOR'S BOWERBIRD

Single maypole surrounded by a circular moss court

SPOTTED BOWERBIRD

Avenue with two courts and wide display area

STREAKED BOWERBIRD

Walled maypole with thatched roof

FAWN-BREASTED BOWERBIRD

Avenue with two courts, all on a raised platform of sticks

SEXY SONGS

Birds use their voices for both courtship and defense. Most songbirds learn to sing from a male tutor during a sensitive phase analogous to the best time for humans to learn a mother tongue. Many species of songbird show regional dialects because, much like human languages, many birdsongs are culturally transmitted. Song sparrows in one part of California sing differently from those in another, even though they are all recognizable as song sparrows.

Backyard Bird

WHITE-CROWNED SPARROWS

The most virtuosic male white-crowned sparrows can pull off songs that are simultaneously rapid and combine a broad range of notes. These high performance songs signal a more serious territorial threat to neighboring males.

Just as with the evolution of elaborate dances or bowers, many bird species have selected for mentally and physically challenging arias. Larks are famous for singing while performing aerial acrobatics. The male skylark sings while ascending in spirals 330 feet (100 meters) in the air. A vocal signal of quality can also involve the ability to sing at a high range of frequencies (high bandwidth) while keeping up a rapid trill rate. This combination is mechanically hard for male birds to accomplish. Female white-crowned and Lincoln's sparrows prefer these high-performance songs, and males use song performance to size up competitors. Sexy song is costly; for example, swamp sparrow males raised with less food have smaller song nuclei, the parts of the brain birds use to learn their songs.

The ultimate examples of mental prowess are open-ended learners that never stop incorporating new elements into their courtship songs. European sedge warblers with the most elaborate songs and the largest repertoire are the first to attract a female when they return from wintering in Africa. Canaries, mockingbirds, and starlings all continue to add new compositions throughout life.

The use of song to attract the opposite sex is not limited to males. Female alpine accentors will often sing to attract multiple males, so as to obtain more paternal care for their chicks. In other species, such as rufous and white wrens, both sexes sing back and forth in duets, sometimes to guard their partners. Red-backed fairywrens, by contrast, often breed in groups, and use both duets and group choruses to jointly defend their territories. Individual fairywrens are more likely to raise their voices and join in if they hear a group member or partner singing.

Bird bills are often perfectly adapted for a particular food item, but evolving a particular bill shape as a tool can also affect birdsong and sexual selection. Crossbills specialize on eating seeds from conifer cones, and such fiddly work has resulted in bills that are best suited to extracting a very particular type of seed. Some specialize on lodgepole pines, others on spruce or larch, and so on, each with the appropriate equipment. Different bill shapes cause calls to differ. Crossbills prefer to mate with similar-sounding individuals, which reduces mixing of genes for each bill type. As a result, few crossbill chicks grow up with intermediate bills that are mediocre for extracting the seeds of any particular conifer.

TOP LEFT

Male superb lyrebirds are legendary vocal mimics that can imitate anything from other species in the Australian bush to chain saws and the clicks of camera shutters.

LEFT

The bills of crossbills cross at the tip and differ in size and shape across subspecies or races, depending on the particular seeds they eat.

HOLDING A TERRITORY

Most of us are more comfortable on our home turf, and fight hard to keep it, especially if we have children and not just ourselves to protect. How hard an individual fights to guard a territory can vary depending on their own condition relative to others, how much they stand to lose in terms of future reproduction, and how well they know their own turf.

BELOW

Blue tit males with crests that reflect more UV light are of higher quality both as fighters and parents, and are much more aggressive toward others with similarly bright crests.

Male dunnocks rise in status up to a peak at three years old, while those that make it to the ripe old age of seven don't try to occupy territories at all, but keep to themselves in gaps between feeding territories. Biologists temporarily kidnapped males, and observed who was able to regain their former territories. Alpha males returned to find either a beta or neighbor lording it over their home and mate. They were least likely to defeat usurpers in their prime, but were also more easily deposed by their subordinate partners

than by neighbors. This is because both the former alpha and beta males knew their territory inside out, giving both an edge over an invading neighbor, but not over each other. In short, partners can often turn out to be enemies from within, and are a lot harder to overcome than comparative strangers.

However, the main factor determining a male dunnock's ability to regain his former position was not age or familiarity with a territory, but how hard he fought. Both monogamous males and alphas fought harder to regain their territory if they had already invested in a family, making the territory more valuable and worth defending than if they were taken away before the female had laid any eggs. There is a "success breeds success" feedback in dunnock contests, because alpha males are best able to monopolize access to a territory and females, meaning they invest more in reproduction, making the territory more valuable, so that they fight harder to keep it, and the harder they fight, the more likely they are to win.

Black-capped chickadees get information on the competitive ability of challengers through eavesdropping on the neighbors' songs. In other birds, visual signals advertise an individual's quality to both competitors and prospective mates. The size of a male house sparrow's black bib matches the quality of territory he occupies. Males with larger bibs also copulate more often both within and outside their pair-bond, sometimes using forced copulations, and have larger testes to make more sperm.

ABOVE

Male long-billed hermits, a common hummingbird in Costa Rica, have evolved dagger-like bills, which they use to stab each other in the throat during contests.

LEFT

Male house sparrows use the size of their black bibs to signal status and quality to females and other males.

SPERM WARS

While males can compete for mates by evolving weapons or showy displays, in species where females commonly mate with more than one male, competition takes place directly between sperm. Both sexes have evolved all sorts of elaborate adaptations in response to escalating sperm armament and to defenses by females to retain the freedom of reproductive choice.

In species where females mate with multiple males, either openly or via EPCs, males have evolved extra sperm production and storage capacities, the better to outcompete the sperm of rival males by swamping them. This comparison is most obvious between closely related species with different mating systems. For instance, fairywren species where males experience more sperm competition have larger testes for their body size. They have also evolved another trick, in the form of a protruding muscular cloacal tip, which appears to stimulate females to be more receptive to their sperm. This tip is longer in species that have relatively large testes and in which females mate with more males.

Males can strategically alter sperm investment. For instance, a rooster increases sperm production when surrounded by more competitors, and invests less and less sperm in mating with the same hen repeatedly. However, his sperm investment is revived by exposure to a new hen, in what biologists call the "Coolidge effect." When visiting a chicken farm, the First Lady, Mrs. Coolidge, was told a rooster could copulate dozens of times a day, to which she reportedly replied, "Tell that to the President." The President then asked if this always involved the same hen, and was told no, to which he replied, "Tell *that* to Mrs. Coolidge!"

LEFT

There can be intense competition among male superb fairywrens and, as a result, they have evolved large testes for sperm storage.

DUNNOCKS

Dunnocks may appear to be dull little garden birds on the surface, but their colorful sex lives have illuminated much of what biologists know about bird mating systems and sexual conflict. Like in most birds, dunnock copulation is fleeting. The pair touch cloacas for the briefest of seconds, and sperm is transferred from male to female. However, males have a rather nasty habit that has disturbed and confused birdwatchers since the early 1900s. When females are ready to copulate, they signal with a raised tail and a flirtatious little wing tremor, as is seen in most birds. Instead of obliging, males peck hard at the presented cloaca for as long as two minutes, until the female had expelled a neat little glob, after which they proceed to mate. Males wait for this droplet and inspect it closely before copulating. In other words, they habitually force their mates to eject sperm from a previous copulation to avoid being cuckolded.

MATING GAMES

Game theory has been co-opted from economics to explain the existence of multiple mating strategies. What should an individual bird do to maximize its reproductive investment given what everyone around him is doing? Some birds have evolved mating strategies they are born with, while others can switch flexibly between strategies to accommodate unpredictable environments.

Multiple male mating strategies

The ruff is a sandpiper in which males, which display elaborate feather collars, gather on leks to win females. Unlike other lekking species, such as grouse or birds-of-paradise, male ruff come in three different genetically fixed flavors. Territorial "independent" males, with resplendent black or chestnut ruffs, are the big bad hunks that most females want to mate with, and the biggest and strongest of these obtain the best territories and the most matings. However, there is also a "satellite" male variant, which sports a snowy-white ruff, and tries to forcibly steal the odd copulation from females gathered around an independent male. Independents tolerate satellites because females are more attracted to larger groups of males and satellites are relatively unsuccessful at mating, so are five times less common than independents.

It wasn't until 2006 that ornithologists discovered a third variant, which they dub "faeder" males, after the Old English word for father. Faeders are sneakers that resemble female ruff and invest almost all their energy into disproportionately large testes rather than external ornaments and fighting. Only 1 percent of males are faeders, suggesting this strategy is only just successful enough to persist at a very low frequency.

ABOVE AND LEFT

Male ruff (above) are named for their breeding plumage, which includes elaborate neck feathers in black or white. However, some males are sneakers, which look almost identical to females (left).

Colorful personalities

Consider a situation in which potential mates advertise their personalities and parenting abilities by the color of their faces. Gouldian finches are endangered songbirds from northern Australia that look like painted children's toys. Their faces are either pure black or crimson, (or occasionally orange, most commonly seen in captive populations). In the wild, birds of one face color tend to mate together, and have more offspring than couples with mismatched face colors. Black-faced finches are about three times more common than red-faced finches, so one of the puzzles is why one type doesn't simply take over.

The answer probably has something to do with the personality types associated with each color. Red-faced finches of both sexes have higher testosterone levels, making them more aggressive and apt to displace the milder black-faced finches from the best nest cavities and water holes. However, this pushy personality comes at a cost, including chronic stress, a poorer immune system, and poor parenting abilities, all of which means that red-faced individuals tend to be shorter-lived and leave fewer offspring than their more placid black-faced counterparts. Both personality types persist because they each have more success in the presence of the other. Red-faced finches do well as a vocal and greedy minority that can take over the best resources from the milder black-faced majority. But when there are too many aggressive red-faced pairs squabbling over mediocre territories and forgetting to feed their offspring, it pays off to be a placid black-faced finch.

RIGHT

Gouldian finches come in two genetically determined colors that are closely tied to personality. Red-faced finches are pushy, while black-faced birds are more placid.

In white-throated sparrows, genes for personality and head stripe color sit on the same chromosome. Individuals with opposite stripe colors (tan, above left; white, above right) tend to pair up.

Opposites attract

Rather like the Gouldian finches, white-throated sparrows from North America come in two colors that are genetically programmed to correlate with a personality type. Individuals with a tan stripe on their heads are gentle, faithful partners and conscientious parents, whereas those with a white head stripe are more belligerent and apt to stray from their social partner.

In contrast to the Gouldian finches, in this species, opposites pair up. A pugnacious white-striped female is better off with a tan-striped stay-at-home dad who remains at the nest while she is busy defending their territory or gallivanting with neighboring males. Tan-striped females also prefer tan-striped males, but are usually outcompeted by white-striped females and forced to settle with white-striped males, which neither female morph prefers. Pairs of the same color are rare and tend to reproduce less. Two white-striped birds tend to neglect their offspring, and two tan-striped birds get pushed around so much that they end up with poor territories.

BIRDS OF A FEATHER

Most of sexual selection theory focuses on explaining highly exaggerated weapons or ornaments that make the sexes so different from one another, but in many birds, sexual selection is more symmetric. When there are benefits to being in a mutually compatible relationship, it can be to the advantage of both sexes to use the same cues to attract and choose a mate.

In many species with long-term monogamy, individuals that are more similar tend to raise more chicks than mismatched pairs, even if one of the individuals is objectively of higher quality, such as being larger. For instance, in brant or brent geese (depending on which side of the Atlantic you live on), larger individuals typically live longer and have more offspring. However, individuals tend to pair for life according to size, and small couples outreproduce couples of very different sizes, which invariably produce the fewest surviving offspring. Similarly, barnacle goose and pinyon jay couples of similar size have the highest reproductive success.

Mutual sexual selection can explain the evolution of ornaments shared by both sexes. Australian black swans choose lifelong mates based on how many curly feathers they have on their wings. Pairs with the most curly feathers occupy the best territories and offer the best resources for their cygnets, which means they tend to raise the most cygnets to adulthood. Western bluebirds prefer to pair up with bluer individuals; females paired with a bluer male are less likely to be unfaithful.

LEFT

King penguins choose one another by the brightness of the yellow and orange decorations on their bills and feathers. As the pigments giving rise to these colors have antioxidant properties, brighter birds may be advertising their better health.

Budgerigar females are most attracted to males that sound similar to themselves, and over the course of courtship, males begin to imitate the voices of females they are wooing. Once egg laying begins (and copulation ends), their voices diverge. Male budgies that have been experimentally brain lesioned are less good at mimicking the females they are courting, and even if they do succeed in getting her to pair, the female engages in more EPCs than if she were paired with a male who was better at learning to sound like her.

Similar personalities often attract, and do better than couples that are very different. Steller's jays, the western American equivalent of the blue jay, are highly adaptable and inventive birds, like most of the crow family. These birds prefer to pair up with individuals of the same personality type,

so shy, risk-averse individuals that travel less tend to pair up, and the more adventurous individuals who take more risks and travel more also prefer one another. The more similar a pair is in personality, the higher its chances of rearing chicks successfully.

Many personality types are caused by baseline hormone levels, such that individuals with similar hormone levels may find it easier to agree upon important decisions such as how far to forage for food, how often to take risks, or how aggressive to be toward intruders. Pairs of greylag geese with similar testosterone levels year after year have more offspring because they are more likely to nest in any given year, produce larger clutches of eggs, and have heavier, better-provisioned eggs that are more likely to hatch into strong chicks that survive to adulthood.

Backyard Bird

GREAT TITS

Both sexes look very similar in the socially monogamous great tits, and there is mutual sexual selection for a larger black breast stripe, yellower breast feathers, and immaculate white cheeks. Females with more extreme ornaments are heavier, have a more robust immune system, and produce higher-quality offspring, while males with a wider chest band are better fighters and providers.

LEFT

Steller's jays are the blue jays of Western North America. These clever corvids are more successful breeders if members of a pair have similar personalities.

FAMILIARITY BREEDS CONTENT

In addition to looking for someone just like you, another way to achieve efficient coordination as parents is to stick with the same individual for a long time, growing to know your partner as well as or better than you know yourself. Birds that mate for life almost invariably need two parents to raise offspring, and there is strong evidence that partners that are more familiar with one another are more efficient and successful at reproducing. Barnacle geese, for example, have a very low divorce rate of about 2 percent, and longer-lasting couples raise more chicks.

The benefits of maintaining long-term pair-bonds must outweigh those of divorce in species like many albatrosses, or great crested grebes, where pairs routinely renew their commitment by going through elaborate and highly ritualized displays every breeding season. These displays become increasingly synchronized over time, and one can only conclude that the outcome of all that dance practice is more coordinated and efficient parenting.

Bearded reedlings, which also pair for life, form their pairs as juveniles, which gives both partners plenty of time to know one another before beginning to breed. By experimentally restricting the length of bearded reedling courtship, biologists have shown that pairs with more time to form a bond are more coordinated in their reproductive efforts. They begin nest building more synchronously and raise more chicks than do pairs that were given less time to bond.

LEFT

Great crested grebe couples present gifts of pondweed to their partners as part of an annual *pas de deux* that renews and strengthens their lifelong social bond.

BELOW

Whooping cranes evolved to emphasize quality over quantity in every reproductive investment, producing relatively few offspring that take a long time to mature, thus putting their numbers at risk.

The importance of choice

Observations of whooping cranes show that these birds begin choosing their lifelong mates early—sometimes before reaching reproductive maturity. Much like young human couples, whooping crane pairs usually spend a year or more in a stable relationship before breeding. Pairs that start getting to know each other earlier, before reaching reproductive maturity, have more time to practice the skills for successful parenting and often end up with higher social status, better territories, and lower stress levels than those who paired up later and don't know their partners as well.

When biologists allow zebra finches to choose their social mate, pairs have 37 percent more offspring than those in experimentally "arranged marriages," where birds had no choice. This is not due to any inherent genetic worth of either individual, nor to females investing more in eggs when they paired with a preferred mate, because the eggs were experimentally switched to separate the effects of maternal investment from the feeding of chicks. Zebra finch pairs that got to choose their partners were simply more compatible at a behavioral level, and managed to feed chicks more because they were a more efficient team than pairs that were forced together.

SEX IN THE CITY

Urbanization and human development tend to be bad for the health of most birds, and can disrupt courtship behavior. City birds have higher levels of chronic stress and urban house sparrows even have higher cholesterol levels from eating discarded fast food. Noise from oil wells and roads elevates the stress levels of greater sage grouse, and disrupts their breeding behavior in numerous ways.

The impact of noise pollution on birdsong has led to urban birds adapting in a variety of ways. In Berlin, nightingales raise their voices in response to the noise of traffic. They even sing more loudly on weekdays than on weekends. European robins in cities sing more at night, not just because of the city lights, but because there is less noise to interfere with their songs than during the day.

Some birds don't just adjust the volume or timing of their songs; they also sing at a higher pitch, to be more easily heard above the low-pitched background noise typical of cities. Great tits, North American song sparrows, and house finches in urban areas all sing at a higher pitch than their country counterparts. Even more impressive, individual male great tits flexibly adjust the pitch of their songs in response to background noise.

In other species, there is evidence that city life is causing some reproductive segregation within species. On San Diego university campus in the early 1980s, a population of dark-eyed juncos began to stay all year round instead of migrating back to breed in the mountains. These birds have since diverged significantly from the migratory juncos, including in sexually selected traits such

ABOVE

European nightingales in cities have adapted to the noise pollution from traffic by raising their voices.

Male collared flycatchers are responding to a warming climate by evolving smaller, less attractive white patches on their foreheads.

amount of white on their tails. Males in other populations use their tails in posturing during territorial disputes, but these year-round residents know their neighbors so well that no one bothers much with display and posturing, so have no need for the white badges of dominance.

More generally, human activities can shape the evolution of bird courtship through more global effects such as climate change. Collared flycatcher males have a snowy-white patch on their foreheads. Males with larger white patches tend to win more battles and more mates, and are better parents. However, whiter males are less likely to survive after a warm breeding season, so there has been a distinct decrease in the fitness of whiter males over the last 35 years. Warming temperatures have caused the survival benefits associated with less white to increasingly outweigh the reproductive benefits of having more white on one's forehead.

THREE

NESTS & EGGS

From male megapodes that build a compost heap to incubate their eggs and use their tongues as a thermometer, to the intricate nests of weaverbirds (left), birds have evolved countless ways to keep their eggs safe. What accounts for the variety of nests, eggs, and incubation behaviors among birds?

NEST ARCHITECTURE

Most avian architecture serves the primary purpose of providing a place in which to rear a family. Nest design reflects both the materials available and the function of a nest.

Bird nests range from nothing but a bare branch on which white terns lay their unshielded egg, to the mud and stick monstrosities that hamerkops construct. These African waterbirds are barely taller than a cocker spaniel, and weigh about a pound (0.45 kg). Yet they are such compulsive builders that a pair will often make between three and five nests in a year, regardless of their breeding status. Each nest is strong enough to support a person, and is so large that other birds, like sparrows, often nest in its thick walls.

Construction materials and techniques

Silk from spiderwebs or butterfly and moth cocoons is the perfect combination of strong, sticky, and elastic for birds to use in nest construction. Hummingbirds use silk to anchor their nests to branches. The silk, alternated with bits of moss

TOP AND ABOVE

Hamerkop nests (top) can be 4 ft (1.5 m) wide, while white terns (above) require just a branch.

RIGHT

Female Anna's hummingbirds take about a week to make a nest of fluff bound together with silk, and decorated with flakes of lichen or even paint chips, which are sometimes stolen from nearby nests.

or lichen, also serves to form a type of adhesive, enabling hummingbirds to form soft, flexible, but resilient cup nests half the size of a walnut.

Members of the Cisticolidae family, tiny songbirds found across Asia and Africa, use silk in even more ingenious ways. The rattling cisticola sticks growing grass together with concentric circles of silk on the inside of a nest that ends up looking like a narrow vase. Tawny-flanked prinias weave a little pouch from plant fibers and stitch it to green leaves, offering both support and camouflage.

Other birds use long fibers to construct nests. Weaverbirds and many New World blackbirds weave structures that resemble intricate baskets. Although these weaving skills are innate, individual birds improve with practice. A Baltimore oriole nest can contain 10,000 stitches.

Mud is another popular building material. White-winged chough and magpie lark nests resemble bowls from a potter's wheel. Most swallow and swift species construct mud cups that adhere to the sides of cliffs or buildings.

Bird hotels

Edible-nest swiftlets construct nests out of a special gluey saliva, which is a Chinese delicacy. Male swiftlets perform all the construction, and their salivary glands enlarge seasonally for the purpose of making these nests. In the 1990s, the Indonesians constructed concrete caves for these birds, to make it easier to collect edible nests than from the caves where the birds naturally nest in their thousands.

Backyard Bird

COMMON TAILORBIRDS

The aptly named common tailorbird, immortalized in *The Jungle Book*, pokes holes along two leaf edges, and stitches them together with silk to form a hanging pouch, which is then lined with soft down, hair, and cotton.

RIGHT

The saliva nests of edible-nest swiftlets have a gelatinous texture which is cooked in a sweet soup as a highly prized Chinese delicacy.

SHELTER FROM THE ELEMENTS

Just as adobe dwellings in deserts or the steeply sloped roofs on alpine chalets are adapted to fit the environment in which they are built, nest designs reflect the habits and habitats of the birds that build them. Nest location, architecture, and construction vary depending on the environmental, developmental, and social pressures different birds face.

Enclosed nests can help to buffer eggs from extreme temperatures. In contrast to the classic cup, many birds enclose their nests with a variety of roofing styles. Eurasian wrens (*Troglodytes troglodytes*, or "cave dweller") are named for their domed nests that resemble tiny caves. Comparing over 300 songbird species, biologists have found that across latitudes, smaller species, like wrens, tend to make enclosed rather than open cup nests. This could be because smaller birds are likely to lose heat fastest in an open nest. Furthermore, species with enclosed nests spend less time incubating, and their chicks devote less energy to staying warm, enabling them to grow faster. Taken together, this suggests that enclosed nests function largely to keep eggs warm. There is no evidence that enclosed nests offered greater protection from predators.

Another superb way to maintain constant temperatures is to nest in a burrow or tree hollow. Some birds, including kingfishers, bee-eaters, and some seabirds, dig tunnels in the soil, whereas others, like woodpeckers and barbets, excavate holes in trees. Gila woodpeckers tunnel into cacti months before breeding, so that the burrow interior has time to dry out before they lay eggs. Many other hole nesters, such as bluebirds, tits, chickadees, and wood ducks, usually don't create their own holes, but use existing cavities, including old woodpecker nest holes.

LEFT

Red-bellied woodpeckers often nest repeatedly in the same dead tree, but typically excavate a new nest cavity below the previous year's.

RIGHT

The rufous hornero (horno means oven in Spanish) uses its bill to construct a mud nest five times its own body weight.

PROTECTION FROM PREDATORS AND PARASITES

Nest design often encompasses a variety of antipredator features. Just as castles had moats to deter intruders, many waterbirds take advantage of their habitat to deter terrestrial predators. Some grebes construct floating nests. Jacanas, also known as "lily trotters" or "Jesus birds" because their long toes enable them to walk on lily pads, construct nests on mats of floating vegetation. Eurasian coots construct such a substantial and sturdy island on which to nest that it can support a sitting person.

ABOVE

The shiny, wet-looking surface of a jacana's eggs help them blend in to their surroundings on a floating nest.

Antipredator defenses

Even cup nests or scrapes are often camouflaged with surrounding materials that make them difficult to detect. Horned larks nest on the open tundra, and camouflage their cup nests with lichen and a scattering of stones. Terns litter their nest scrapes with bits of shell and pebbles, breaking up the nest outline. Hummingbird nests are so covered with flakes of lichen that they often resemble small bumps on a tree branch.

Little grebes and yellow rails cover their eggs with vegetation when they leave the nest to feed, which helps hide the eggs from predators and maintain the right temperature and humidity. Virginia rails make several dummy nests, possibly to confuse predators, and one of their relatives, the purple swamphen, invariably lays eggs in the least conspicuous of their collection of nests. Some wrens, waxbills, and weaverbirds construct nests with false entrances. Red-cockaded woodpeckers, which nest in pines, deter predatory snakes by stripping bark, resulting in a sticky resin that extends for several feet around the nest entrance.

Rock wrens use up to 3 lbs (1.4 kg) of stones to pave the entrances to their nests. Each stone can weigh more than half the bird's body weight. These stones help reduce flooding, and act as a predator alarm system, because they are so loosely arranged that anything moving over them creates a racket that the wren inside its nest hole can hear. Blackstarts in the Middle East appear to have converged on the same use of stone pavements as a predator alarm system.

LEFT

Purple swamphens make multiple nests, and only lay their eggs in the best concealed nest.

LEFT

Tree hollows with narrow entrances can be safer than building a nest in the open, but they come at a cost. Female hornbills seal themselves in with a mixture of mud, droppings, and fruit pulp, leaving a narrow slit through which the male can deliver food. A female also plucks out all of her flight feathers in order to fit inside the tree cavity, making her entirely dependent on her mate.

Narina trogons and hoopoes also nest in holes, and both species are armed with liquid predator deterrents. The trogons shoot an evil-smelling liquid from their preen glands, and hoopoe chicks shoot streams of poo at any predator that pokes its head into the nest opening. Similarly, cape petrels and fulmars vomit foul-smelling stomach oils when defending their nest burrows.

In addition to mobbing or alarming, some birds perform distraction displays that lead approaching predators away from a nest. The most famous of these is the broken-wing display of killdeer plovers, where parents flap about as if injured and quickly scurry away only when they've lured a predator far from their eggs or chicks. Similar deceptions have evolved in other species. The red-billed leiothrix, a colorful and highly popular caged songbird in East Asia, uses a noisy wing-flutter display to lure predators from its nest. Wilson's phalaropes have a coordinated response to predators. The female, who contributes almost no other parental care, will sit on the nest briefly to hide the eggs, while the male performs a distraction display.

Neighborhood watch

More commonly, other birds serve as the best alarm systems. White-winged choughs nest near Australian kestrels to take advantage of the falcon's greater vigilance against predators.

There is also safety in the presence of strong neighbors, and snow geese like nesting near snowy owls. Common waxbills in Africa simulate the presence of dangerous predators by regularly replenishing their nests with fresh carnivore scat. Using scat supplementation and removal experiments, biologists have shown that cat scat deters rodents from eating waxbill eggs. Black larks from the Eurasian steppes pave the area around their ground nest with livestock dung. Experiments revealed this had no effect on predation rates, but

TOP

Distraction techniques are employed by species such as the red-billed leiothrix.

ABOVE

Great gray shrikes are famous for impaling their prey on thorns. Pairs are highly territorial and have loud alarm calls. They join forces with other songbirds such as thrushes to defend their nests.

American robin parents maintain a hygienic nest by removing the fecal sacs produced by their offspring.

biologists have hypothesized that the dung does help deter grazing livestock, thereby lowering the risk of hooves trampling the nest.

Similarly, burrowing owls of the American west often line their burrow entrances with livestock manure, and occasionally grass, paper, or plastic. There is evidence that manure attracts prey such as dung beetles, but the function of other materials is unclear.

Good housekeeping

Birds must also keep their nests clean, partly because a rank nest attracts more predators and parasites. Many chicks, from blue jays to nuthatches, produce neat little fecal sacs that their parents ferry out of the nest as they fly away to collect food. Others, like hornbill or hawk chicks, aim their projectile poo so that it lands outside the nest altogether. More fascinating are the antimicrobial substances some birds employ in their nests. Biologists were puzzled to find collections of cigarette butts in house sparrow and house finch nests, but then noticed that nests with more cigarette butts contain fewer lice.

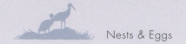
LOCATION, LOCATION, LOCATION

As any property owner knows, location can be of the essence when it comes to determining the value of a home. Some preferred nesting sites are so scarce that birds will reuse them long beyond the lifetime of an individual.

Eastern phoebes often nest in barns in New England, and many people have observed the same mud nest being used for decades, even though individuals live no longer than 10 years. Cliffs are superbly safe places to nest and are especially favored by eagles and falcons. By carbon-dating the guano that can be up to 6.5 feet (2 meters) deep on a cliff ledge, scientists have found that some gyrfalcon sites have been used for 2,500 years. Similar guano dating methods suggest that some Adélie penguin colonies have been used for over 40,000 years.

Although most birds incubate their eggs using the warmth from their own bodies, some species use more creative means. The maleo, a curious and distant chicken relative found only on the Indonesian island of Sulawesi, buries its eggs in the sand, allowing geothermal heat and the sun to incubate them. Desert larks nest under rocks or bushes for shade, but orient their nests to catch the morning sun, to warm them up after the cold desert nights.

Birds also choose their nesting location to minimize predation. Japanese quail choose nest sites where their eggs will be hard for predators to spot. In an experiment where biologists presented quail with an array of different colored sand patches, females whose eggs were mostly pale and creamy chose to lay on sand that best matched the background color of their egg. In contrast, females whose eggs had lots of large, dark splotches preferred darker sand, which breaks up the egg's outline. Female baya weavers of Asia choose males with nests in the safest locations, preferring them to be higher up, in thorny trees, and on the ends of the thinnest branches.

TOP LEFT

Gathering guano to carbon date a gyrfalcon nest.

LEFT

Female red-winged blackbirds prefer to nest above water, so as to discourage land predators like raccoons.

COLONIAL NESTING

Massing together can provide safety in numbers, but also comes at a cost. Colonial living can exist because of a net benefit to individuals, even if it is suboptimal for the group.

Although living in a massive group dilutes the chances of any individual being eaten, a large colony also attracts predators. Skuas and gulls often hang about the colonies of smaller marine birds like puffins, petrels, or penguins. California gulls are even known to prey on the eggs and chicks of other colony members.

In addition, building materials can be in short supply, making it tempting to help oneself from a neighboring nest. Even before laying any eggs, anhinga females stay on their nests and are fed by their mates, to prevent other colony members from pinching sticks. Sand martins, known as bank swallows in America, live in colonies of up to 30,000 pairs, all burrowing into the side of a sandy bank. The colonies sometimes get so large that they can cause an entire section of bank to collapse from the collective weakening by thousands of burrows.

Sociable weavers of southern Africa nest in vast straw condominiums comprising the conjoined nests of as many as 100 families. New apartments are added from the bottom, and the entire structure requires year-round repairs. The structures can be used for over a century, growing so large that trees can collapse under their weight.

LEFT

Male sandmartins begin each burrow, which they advertise to females. If a female approves, the male completes the burrow and the female makes a nest inside.

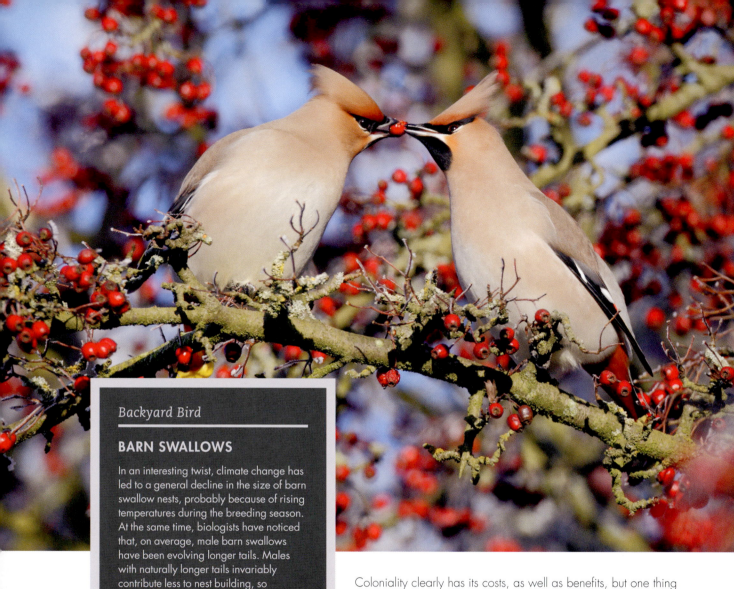

BARN SWALLOWS

In an interesting twist, climate change has led to a general decline in the size of barn swallow nests, probably because of rising temperatures during the breeding season. At the same time, biologists have noticed that, on average, male barn swallows have been evolving longer tails. Males with naturally longer tails invariably contribute less to nest building, so warming temperatures have allowed sexual selection by females to increase the average tail length of male swallows at almost no cost to chick survival in smaller, cooler nests.

Coloniality clearly has its costs, as well as benefits, but one thing many colonial species have in common is a food source that is impossible or unnecessary to defend, reducing competition. Swallows and swifts hawk for flying insects, and most marine birds hunt away from their nests, making it unnecessary to defend a territory that encompasses all the food for raising a family. Similarly, Bohemian waxwings rely on an unpredictable food source of berries, so are not territorial, and tend to breed nomadically wherever there is plenty of food, rather than returning to the same place year after year like most other songbirds.

ABOVE

Bohemain waxwings are aptly named because their unpredictable food source of berries leads to a far more nomadic lifestyle than most birds.

NESTS AS SIGNALS

Extravagant architectural gestures are not limited to human creations like palaces or even McMansions. In many birds, nests and their accompanying decorations function as signals, both of an individual's quality as a partner, and of a pair's social standing.

Black kites line their large, open nests with conspicuous white streamers from any material they can gather, which often includes a lot of plastic. The more white in a nest, the fewer territorial intrusions a pair encounters, suggesting the white functions as an effective "no trespassing" sign to other black kites. Older, more dominant couples fly more white flags.

BELOW

A black kite nest flying white "flags" as signals of superior status.

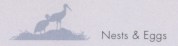

ABOVE

White storks often nest on buildings, as this pair is doing in Marrakesh.

Feathers and plants are also used as signals. Male house sparrows call to their mates when they bring feathers in, as if to say "look what I found for our nest." In response, females presented with more feathers lay more eggs, and feed the chicks more often. Similarly, female blue tits do most of the nest construction, but males will carry feathers in to line the nest. When biologists added feathers to nests to simulate presents from a different individual, males behaved as though they had been cuckolded by reducing their share of parental care.

Having a male provide a nest as part of his courtship is a good way for females to guarantee some paternal contribution before mating takes place. White storks prefer males who are able to win and defend larger, older nests. Brand-new stork nests are small and less attractive, whereas venerable mansions grow in value and size, as breeding couples continually add extensions over the years. The largest, oldest nests typically belong to older, more experienced males who can fight to retain them, and who also tend to rear more chicks successfully.

Male marsh wrens make multiple nests and escort prospective mates through their territory like a real estate agent with a new client. If a female likes one of the nests, she completes the internal decor by lining it with soft feathers. Great egret males begin nest construction, and if their nest attracts a female, she helps to finish it. The function of nests as sexually selected signals could explain why great egrets seldom reuse nests, in spite of the time and effort they take to construct. Female egrets are attracted to good builders.

Black-billed magpie nests look like large, domed balls of sticks, and are mostly built by the male, while the female lines the interior with mud and incubates the eggs. The larger the nest, the more eggs a female is likely to lay, because males that build larger nests tend to be better parents, delivering more food to their hungry chicks and partners. Even great spotted cuckoos, brood parasites that rely on magpies to raise their chicks for them, know about this signal, and prefer to lay in larger magpie nests.

Birds in which both sexes care for the young often use nest construction as part of their courtship and subsequent pair-bonding rituals. Black skimmer courtship involves protracted mutual sand kicking, which helps to make a scrape in which the female lays her eggs. Males tend to kick more sand than females. Chinstrap and Adélie penguins construct stone mounds on which to lay their eggs, to reduce the chances of snowmelt flooding the nest. These penguins will present their partners with stones whenever they greet one another, but stones are in short supply, which leads to a great deal of stealing and guarding of these precious items.

ABOVE

Larger black-billed magpie nests signal that the male that made them is a better builder and parent than other males.

LEFT

Black wheatear males weigh less than a golf ball, but carry up to 50 times their body weight of stones to their nests. A larger stone pile induces the female to lay eggs earlier.

81

EGGSHELL DESIGN

Eggshell textures, colors, and patterns can be used to reduce detection by predators or even increase incubation effort by males.

ABOVE

There is evidence that pigments can lend structural support to eggshells.

All birds that lay blue or green eggs do so with a pigment called biliverdin, which is also an antioxidant. As a result, we can infer that females pay a cost by depositing the pigment in eggshells rather than using it to combat aging. As with other costly substances, blue pigments could be used as honest signals of quality, to induce partners to invest more in joint ventures such as raising a family. Supporting this idea, younger, healthier pied flycatchers lay bluer eggs than older, less healthy females, and males feed their chicks more if they hatched out of bluer eggs. There is also a weak pattern across over 150 songbirds that blue-green eggs are disproportionately common among species with a longer nestling stage (requiring more care) and more polygyny (males might be helping multiple mates).

Protective shells and signatures

Many species have evolved a variety of egg surfaces for various purposes. Red-winged blackbirds that live at higher altitudes have adapted to the dryer air by laying eggs with smaller pores to reduce the chances of the embryos drying out. Biologists measuring thousands of eggs have found that more ultraviolet light penetrates whiter shells and that birds that nest in cavities, or sit on the eggs almost constantly, are more likely to lay white eggs than those with open nests. This suggests that some birds use pigmented shells as sunscreen to protect developing embryos from ultraviolet rays.

Fossils of oviraptor dinosaur eggs from China, dating back to the Cretaceous, also show signs of blue biliverdin pigments. Their nests resemble those of ratites (ostriches and their relatives), which contain the eggs from multiple females, all incubated by a single male. Paleontologists speculate that dinosaur females could have been under selection to lay blue eggs to attract male investment in incubation well before the evolution of modern birds.

AMERICAN ROBINS

Selection by male birds, who provide more care to young from bluer eggs, is currently the most popular explanation for why females of some species have evolved to lay blue eggs. Biologists have tested this idea in American robins, by replacing their eggs with artificial eggs that varied in how blue they were, and then hatching the real eggs in incubators. When the chicks hatched, they were randomly assigned to different nests, to remove any effect of innate quality in either parents or chicks. Male robins that saw their mates incubating the bluer eggs worked harder on feeding the chicks than males whose clutch had been replaced by less blue eggs.

Hoopoes nest in burrows, and biologists filmed females using their bills to paint their eggs with a sticky, smelly secretion from their preen gland. Closer inspection of the eggshells revealed that the surfaces have special depressions, not found on the shells of other species, which help hold the secretions. Curious about the potential antibacterial function of this egg anointing, biologists temporarily plugged the glands of some females with plastic tubes, and swabbed the eggs every few days. They found that eggs without the special coating were less likely to hatch, and contained fewer symbiotic bacteria that would otherwise help prevent harmful bacteria from invading the eggs.

Murres are diving seabirds that breed in colonies on the same cliffs for generations, making the nesting ledges sticky with guano. To keep the eggs clean, their shells are covered with a special coating that works rather like the surface of a nonstick pan. The eggs of common murres also have highly distinctive squiggles on their shells, which may function as signatures to help parents identify their own egg in the densely packed colony. Female coots often parasitize each other by laying eggs in the nest of a different pair. To counter this, female coots have evolved to know their own signature egg patterns, and push parasitic eggs to the outer edge of the nest, where they are most likely to be picked off by a predator.

TOP LEFT

Hoopoes coat their eggs in a specialized antimicrobial secretion from their preen glands, which seems to increase the chances of successful hatching.

LEFT

Like many seabirds, common murres breed in colonies on exposed cliffs.

EGGONOMICS

All parents face trade-offs between investing in current reproduction at the expense of future reproduction, and between the quantity and quality of offspring. To maximize their lifetime reproductive returns, female birds can flexibly alter when to lay their eggs, as well as the number and quality of eggs in each clutch.

Coursers, which breed in deserts, can lay a new clutch of eggs quickly if the rains last long enough to provide food for them to rear a second brood. Brent geese can even resorb eggs if conditions in their unpredictable Arctic breeding grounds become unsuitable for nesting.

Young red-billed chough females that lay more eggs die earlier than those that lay fewer eggs as young birds. Siberian jays live for about 20 years, so it is sometimes worthwhile for females not to put all their eggs in a single clutch so as to live to raise another brood. Females nesting in more open habitat where their nests are more visible to predators tend to lay fewer eggs in a single clutch. In addition, females who delay dispersal and spend a year or so with their parents take longer to begin breeding and live longer than those that start families as soon as they reach reproductive maturity.

LEFT

Female Siberian jays strategically alter how many eggs they lay and when they lay them to maximize their lifetime reproductive success.

Female choice through maternal investment

Female birds invest more in eggs that have a better chance of keeping the family line going, whether due to favorable rearing environments, or superior genes. Peahens lay more eggs if mated to a male with a longer tail. Ostriches prefer males with more contrasting black and white feathers and redder necks, and lay heavier eggs after mating with a more attractive male. Similarly, mallards, ring-necked pheasants, and zebra finches lay heavier eggs with larger yolks when mated to sexier males.

Malleefowl resemble turkeys, but have nesting habits unique to their family. Unlike the maleo of Sulawesi, which uses passive heat from the sun or volcanic activity to incubate its eggs, male malleefowl build and tend compost heaps. They test the temperature with their tongues, and pile on more compost if the mound interior is too cool, or kick away some of the compost if the eggs are beginning to cook. Females move between male territories assessing the various mounds, and tend to choose those maintained at the most stable temperatures of between 90 and 95 degrees Fahrenheit (32 and 35 degrees Celsius). They fight over these most popular mounds, and invest more in the best mounds by laying heavier eggs. As males do all the work of tending the mounds and the eggs within, they only allow a female to lay her egg in their mound after copulation. The chicks hatch and dig their way up through the compost unaided, and run off into the forest without needing any parental supervision, so all parental investment takes place up front in these birds.

In species where males experience much stronger sexual selection than females, sons are a much more high-risk, high-reward investment than daughters. Sons that make it to the top get most of the matings and will provide a female with more grandchildren than even the healthiest daughter. In contrast, inferior or even mediocre sons have next to no offspring, and are a much worse bet than a daughter. Mothers in these species can flexibly invest in expensive but high-reward sons when they are likely to produce top-quality males, and hedge their bets by

BELOW

Female ostriches will invest more in eggs that have been fertilized by a more attractive male.

LEFT

The malleefowl of Australia are members of the Megapode family, named for their large feet, which males use to dig and construct compost heaps insulated with a mound of sand for incubating eggs.

producing more daughters when the offspring are more likely to be middling. The kakapo, a rather hapless, flightless, and highly endangered New Zealand parrot, forms leks, in which top males attract the most mates. Conservationists were initially distressed to find their carefully nurtured kakapos in a captive breeding program producing an excess of sons, until they realized that all the supplemental feeding was making females invest in expensive but high-reward sons. When they put the females on a diet, they got them to produce an even sex ratio.

Similar patterns of flexible sex allocation are found in other bird species. When experimenters pair a female collared flycatcher with a male that is a better provider, she produces more sons. Some female songbirds produce more sons when mated to sexier males, because their sons are likely to inherit their father's attractive qualities and produce more grandchildren than a daughter. Female collared flycatchers mated to males with larger patches of white on their foreheads lay more male eggs. Female zebra finches like red legs, and when experimenters augmented males with little red leg bands, their mates laid more eggs, and more of them hatched into sons, in contrast to females mated to males wearing unattractive green or blue leg bands.

BELOW

The kakapo of New Zealand can adaptively alter the sex of its offspring in response to food quality.

TOP AND ABOVE

Although they are members of the same genus, the American robin (top) and its tropical relative the rufous-bellied thrush (above) have very different approaches to reproduction and risk.

The geography of risk

Predators are one of the biggest threats to avian reproductive investments. Birds have evolved both innate and flexible responses to different degrees of predation, depending on the predictability of the risk to their lifetime reproduction.

Across songbird species, adults in temperate climates have about a 50 percent chance of surviving from one year to the next, whereas tropical species often have a higher adult survival rate of about 75 percent. As a result, temperate species tend to invest in large, short-term, higher-risk investments, rather than saving up resources for a future breeding attempt. Comparing closely related songbird species in the Americas shows that, once one controls for ancestry and size, temperate breeders lay an average of four to six eggs compared to the two to three eggs typical of species breeding in the tropics.

Tropical adults also take fewer risks when raising a family, in favor of living to breed in a future year. When biologists simulated risk by playing the sounds of nest predators or mounting a stuffed predator near their nests, temperate birds took more risks by continuing to feed their chicks at high rates, probably because they have less chance of surviving to invest in another brood. In contrast, parents of tropical species responded to the fake predators by visiting their nests less often and taking fewer risks, showing that they value their own survival (and future reproduction) more than do temperate zone birds. Songbirds nesting in colder climates have the added disadvantage of having to take shorter feeding bouts off the nest, lest their eggs freeze. They compensate for these shorter feeding trips by making more of them, which also makes their nests more noticeable to predators because of the increased activity.

The studies mentioned so far focused on continents. It turns out that because islands tend to be more clement and harbor fewer predators than continents at the same latitude, birds breeding on islands have higher survival rates than those on the mainland. This effect is especially pronounced at higher latitudes, where island breeders have much smaller clutches, longer development times, and an overall emphasis on quality over quantity of offspring, compared to mainland birds.

The downside to evolving on islands is that species tend to lose their innate fear of predators, making them ill-equipped to face immigrant humans and their accompaniment of rats, cats, and other mammalian predators. However, biologists have found that on islands where introduced predators have been around for 700 years, New Zealand bellbirds have adapted by spending less time moving on and off their nests, compared to the same species breeding on islands that have never been invaded by predators.

BELOW

Unlike many island bird species, which lack a fear of mammalian predators, New Zealand bellbirds may be adapting fast enough to avoid extinction caused by introduced cats and rats.

WHO GETS LEFT HOLDING THE EGGS?

Incubation, like gestation, is expensive, requiring time and energy that might be spent on wooing new mates or taking care of oneself to live longer. The males of many bird species assist with or assume the majority of incubation for a variety of reasons, and some do so with flexibility. For some species, it is worth switching between incubation and trying to attract new mates, depending on the relative costs and benefits of each activity.

Males that are more likely to have fathered a clutch of eggs have a greater incentive to invest in incubation. Using DNA fingerprinting, biologists have found that across 16 species of socially monogamous songbirds, females of the most faithful species get the most help with incubation.

Backyard Bird

EUROPEAN STARLINGS

When biologists supplement an area with extra nest boxes, male European starlings spend more time serenading prospective females and less time incubating their current brood. They also spend more time lining nearby nest boxes and singing if biologists remove some of their eggs; if their current family is smaller and worth less than two families, their time is better spent trying to attract another mate. Conversely, when biologists add extra eggs to a clutch, it is more worthwhile for males to incubate the current clutch than to go back on the dating market.

Penduline tit males make exquisite hanging nests from soft material like spiderwebs, plant down, and wool. Then females who approve help to finish the nests off. Eurasian penduline tits are the best studied species in this group, and are excellent examples of sexual conflict over parental care. In this species, both sexes have more offspring in a breeding season if they desert their partner and remate, but this is only the case if the abandoned mate is willing to stay behind as a single parent. When both parents desert, which happens in 30–40 percent of nests, neither sex benefits because none of the eggs hatch.

More recently, biologists found that the closely related Chinese penduline tit experiences less sexual conflict, with half the frequency of mutual desertion found in the Eurasian species. In the Chinese birds, females are more often the ones left caring for the young, possibly because the parents only desert after a female has completed her

ABOVE

Eurasian penduline tits of both sexes stand to gain an evolutionary advantage by being the first to desert the nest, but lose out if both parents desert.

clutch, and there is no difference in the number of eggs reared by a single female or male. In contrast, female Eurasian penduline tits stop laying eggs the moment their mate deserts, so it costs them less to desert too and start again. Penduline tits leave at the start of incubation, much earlier than other birds in which parents are engaged in a tug-of-war over who gets to leave first. Biologists think this is because their nests are so well constructed that nest failure, either due to predation or cold eggs, is relatively low, giving single parents a much better chance of rearing chicks than in most bird species.

At the other extreme, in some species, both parents are so necessary for chicks to survive that parental cooperation, rather than conflict, is the rule. Laysan albatross eggs are so expensive that females can only afford to lay one a year, and two adults are needed to incubate a single egg successfully. Since the early 2000s, biologists had been puzzled to see several albatross pairs on the Hawaiian island of Oahu with two eggs. Upon DNA testing (the sexes are so similar one can't tell them apart just by looking), they discovered that about a third of the albatross pairs consisted of two adult females. Although these females have never been observed copulating, they function like a strongly bonded albatross male–female pair, complete with courtship rituals and the equal sharing of parenting duties.

The strongest explanation for same-sex parenting in this particular population is that the females are making the best of a bad job. For unknown reasons, Oahu has twice as many adult female as male albatrosses. The females seem to copulate with males before returning to the nesting grounds, but due to the shortage of male partners, team up with a female partner. These female partnerships can last as long as 20 years, just like a heterosexual partnership, but females suffer reproductive costs to being in same-sex pairs. Firstly, the pair can only afford to incubate one of their two eggs, so one of the females forgoes reproducing every year. We don't yet know if partners take turns raising each other's young every other year, or if partnered females are usually related so that they still contribute some genes to the next generation by helping raise a niece or nephew, or if one female is dominant over the other and reproduces more.

Secondly, males usually take over the initial three-week long incubation shift, allowing their mate to replenish the energy she invested in their egg. Female pairs do not have this luxury, so one of them has to do double duty, laying and incubating an egg without a break. Probably because this extra work takes its toll on at least one of the partners, females in same-sex pairs raise fewer chicks and die earlier than females partnered with a male.

LEFT

Female Laysan albatross preening each other.

Emperor penguins are the epitome of paternal devotion, with males incubating an egg on their feet for months throughout the long Antarctic winter before their mate returns to relieve them of the hatchling.

FOUR

RAISING CHICKS

Raising offspring is an expensive business, involving lots of food, and sometimes lessons on singing or navigation. What is involved in parenting, and why do some birds routinely have more young than they can afford, leading to infanticide or siblicide to cull the weak or unappealing?

DEGREES OF DEPENDENCE

So far in this book, the drama of cooperation and conflict on the bird family stage has been enacted between parents. However, once chicks hatch, they can directly influence adult investment, leading to conflicts between parents and offspring, as well as between chicks in a nest.

Some young, like those of crocodiles or malleefowl, are precocial: self-sufficient from the time they hatch and dig their way out of the nest. At the other extreme, altricial offspring, like humans or songbirds, are born helpless, completely dependent on adults for food and protection.

Many bird species fall in between the two ends of this spectrum. Ducklings and shorebird chicks are able to find their own food from the moment they hatch, but still rely on their parents for protection. The chicken family, including grouse, turkeys, and pheasants, show their slightly more dependent chicks how to find food. Young rails, including coots and moorhens, can swim but are initially fed by their parents, and young grebes often get to ride on their parents' backs until they get older. Semi-precocial gull and tern chicks hatch with eyes open and the capacity to walk or even swim, but remain at the nest to be fed by their parents. Birds of prey hatch covered in soft down, and some even have their eyes open, but they are still largely altricial because they would be utterly incapable of feeding themselves or escaping from danger.

The more helpless and altricial the chicks, the more adult supervision they need. While most precocial species, like many shorebirds or chickens, can succeed with single parents, semi-precocial or altricial chicks usually need

ABOVE

Common moorhen chicks are less precocial than ducklings, because although they can swim, they are still dependent on their parents for food.

THE GRADIENT OF PARENTAL CARE FROM MOST TO LEAST PRECOCIAL

MOST PRECOCIAL	No parental care.	Megapodes
PRECOCIAL	Protection from parents but can find own food, care by 1-2 parents.	Ducks, shorebirds
	Shown food by mother.	Chickens
SEMI-PRECOCIAL	Hatch with eyes open, covered in down, mobile soon after hatching, but stay in nest and fed by both parents.	Gulls, terns
SEMI-ALTRICIAL	Hatch with eyes open, covered in down, but incapable of leaving nest soon after hatching, fed by both parents.	Herons, egrets, hawks
	Hatch with eyes closed, covered in down, incapable of leaving nest soon after hatching, fed by both parents.	Owls
ALTRICIAL	Hatch with eyes closed, no down, incapable of leaving nest, fed by both parents.	Songbirds

BELOW

Barnacle goose hatchlings have to rely on their fluffy feathers to parachute down 400 ft (120 m) of sheer cliff to find food.

the care of at least two adults. This dependence after hatching poses an investment trade-off for parents. Species with precocial young have to invest disproportionately more energy in eggs and incubation, so that their young are more developed by the time they hatch. If a predator approaches, some of these precocial young can escape by running, swimming, or even flying out of the way. In contrast, altricial chicks require less of an up-front investment in eggs, but parents are more likely to have an entire brood gobbled up in a single mouthful.

LEFT

Songbird chicks like these robins are among the most altricial, depending on their parents for all their needs in the first few days after hatching.

ARE YOU MY MOTHER?

Imprinting is the process by which very young animals learn to identify their parents. Goslings and ducklings are famous for imprinting on the first large object they encounter, waddling behind other waterfowl species, chickens, or even biologists, under the impression that these are their parents.

This can lead to mistaken identities later in life, but because young waterfowl are so independent, misimprinting is seldom immediately costly, and can be quite common.

In particular, many duck species frequently lay their eggs in the nests of other females. In some species, such as wood ducks or goldeneye, this brood parasitism often occurs because suitable tree holes or lakes are in short supply, so the winners of territorial battles end up taking care of the losers' offspring. However, because the ducklings hatch able to swim and feed themselves, the burden of a few extra ducklings is relatively slight, and can benefit their caretakers by lowering the odds that their own offspring will be picked off by predators. But misimprinted ducklings raised by a different species are essentially an evolutionary dead end, as they grow up trying to breed with the wrong species.

Endangered species like whooping cranes and California condors have benefitted from our understanding of imprinting. Captive breeding programs take pains to avoid having their young charges misimprint on their human keepers. From the moment they hatch in incubators, young cranes and condors are fed with hand puppets that mimic their biological parents. Whooping crane keepers even dress up in white suits and have to avoid making any human sounds lest their young charges learn to identify with the wrong species.

TOP LEFT

Konrad Lorenz being followed by imprinted greylag goslings. Greylag geese are the poster goslings of behavioral imprinting, but they aren't as naive as one might suppose. Orphaned greylag goslings choose to join dominant families over those lower in the pecking order.

LEFT

Young whooping cranes are reared by costumed humans to stop them from misimprinting.

PROVIDING FOOD

Birds like the city pigeon can breed all year round, but most birds with dependent young time their breeding to coincide with a seasonal glut of food. Climate change has caused significant mistiming problems for many migratory species, from songbirds to shorebirds that breed at high latitudes.

Backyard Bird

WOODPIGEONS

The pigeon family are unique in feeding their altricial chicks a milky liquid from their crops. The same hormonal pathways involving prolactin evolved independently to induce milk production and feeding in both pigeons and mammals.

Great tits have evolved to time their breeding to coincide with a large caterpillar emergence when their chicks are at their most demanding age. Climate change has been causing the caterpillars to emerge earlier, so that many bird families miss the peak in food abundance. This mismatch is causing a decline in historically common songbird populations.

BELOW

Albatross parents feed their chicks a mixture of semi-digested seafood gathered on fishing trips. Plastics have accumulated in the oceans in such quantities that they can constitute a high proportion of the food transferred from albatross parent to chick, damaging their delicate guts and killing them.

Specializing in an unusual food source or a harsh climate has the benefit of minimizing competition, but it also helps to be able to adapt rapidly to unexpected changes. Snail kite chicks are intensive to raise because of their highly specialized diet of snails. These birds are now endangered in Florida because development has drained most of their habitat, causing a dearth of the snails they rely on for food.

In contrast, the corvid family, which includes crows, magpies, and ravens, is renowned for its intelligence and resourcefulness. Species that live in harsh climates, such as Clark's nutcrackers, make up for a dearth of food at the start of their breeding season by storing seeds with which to feed their chicks. An adult can remember tens of thousands of seed cache locations. Gray jays begin nesting in the late winter or early spring, when temperatures are often below freezing. These corvids also amass food stores before winter sets in, and often glue their food under bark or in crevices with sticky saliva.

BELOW

Snail kites have curved bills adapted for extracting apple snails from their shells. Relying on such a specific diet puts survival at risk when that food source is under threat. .

PARENTAL COOPERATION AND CONFLICT

Parents of all but the most precocial young benefit by teaming up to provide chick care. However, some parental teams are more coordinated than others. This variation is both a function of necessity and individual circumstance.

Songbird hatchlings are so needy that two adults do a much better job of providing enough food and warmth than a single parent. Biologists have shown that removing one parent from song and seaside sparrow families led to half the chicks being raised compared to families that were left intact. Similarly, when they glued weights onto the tails of European starlings, making them less efficient at feeding the chicks, their unhampered partner attempted to compensate but was unable to do so fully, and they raised fewer chicks than a fully functional pair.

Parental cooperation

Seamless cooperation is more common between parents with equal long-term shares in one another's reproductive output. Zebra finch couples stick together for years, so both partners stand to lose if their mate gets too worn out to successfully raise future broods. Cockatiels, another species that remains monogamous for years, benefit from maintaining a strong relationship. Couples that stuck together more during the nonbreeding season, engaged in more mutual preening, and fought less were more coordinated when it came to incubation and chick feeding. Ultimately, compatible long-term couples benefit because they raise more chicks, not just within a season, but also across their lifetimes.

TOP RIGHT

Blackcap parents coordinate feeding trips to minimize the amount of movement that could attract predators to their nest.

RIGHT

In spite of a high divorce rate of more than 40%, Australasian gannets are more likely to reproduce successfully with a long-term mate than with a new partner.

Parental conflict

However, all is not sweetness and light among bird couples, because of the twin specters of desertion and adultery. Although 90 percent of birds are socially monogamous, most partners only stay together for a single breeding attempt, and have no reproductive interests in the longevity or future breeding success of their mate. Furthermore, because most of these species engage in extra-pair copulations, it often doesn't pay a male to care for a brood in which he has only a partial genetic stake. These situations lead to considerable conflicts over resources within families because not everyone is equally related to everyone else.

Males can use a rule of thumb like how much time they spend guarding their mate to assess their certainty of paternity. When biologists kidnap a female collared flycatcher for just an hour during her fertile period, her mate thinks she is off engaging in EPCs, and reduces the amount of care to the chicks that subsequently hatch. This flexible response makes a lot of sense, because a male's genes gain no benefit from him wasting time and energy caring for a brood that aren't descended from him.

Conversely, females can use matings to increase the amount of help they get raising the offspring. American crows that mate with more males in their group get more help feeding the chicks, because those males then have a genetic stake in the brood.

Testosterone influences how males can switch between being sexier, to win more matings, and being better fathers, investing more in caring for fewer children. Junco males with higher testosterone levels attract more females for extra-pair copulations. More testosterone doesn't increase the number of affairs male house sparrows or blue tits have, but they attract more mates and provide less paternal care.

TOP RIGHT

Collared flycatcher males tend to set up second homes not far from their first. Females are often deceived into mating with an already mated male who will desert her to help raise his first family, leaving her to raise 40% fewer chicks than if she had help.

RIGHT

Older male yellow warblers have more red on their breasts and are more attractive to females, siring more extra-pair young, but they also contribute less to feeding the chicks.

INFANTICIDE

Chick killing is the most evolutionarily expensive way to wage parental warfare, because it is much harder to start afresh, and parents lose a larger investment than if the nests or eggs were destroyed.

Many disputes and infanticides are ultimately due to competition for limited resources. Males that take over a territory or nest often kill all the previous male's offspring. This behavior makes evolutionary sense for the incoming male, because he has no genetic stake in the current brood, and it would cost him valuable time to let a female raise them to independence, even if he didn't help with the childcare. Similarly, male common loons compete over a limited number of lakes, and an invading male usually kills the previous male's chicks so he has a chance of raising his own family before the short northern breeding season ends.

BELOW

Common loon chicks can swim and dive underwater 2–3 days after hatching, but continue to ride on their parents' backs, and remain dependent on their parents for food.

Conflict between mated pairs over limited resources can also lead to infanticide. Like most parrots, crimson rosellas and green-rumped parrotlets nest in holes, and suitable holes are often in short supply. In both species, birds will destroy the eggs of other pairs in competition for the best holes. Holes with more cover are safer from predators, and pairs who live in them face more harassment and infanticidal attacks from their neighbors than pairs with less coveted nests.

Daughters vs. sons

Perhaps the most disturbing form of infanticide in birds occurs when parents kill their own offspring to have more daughters or sons. Eclectus parrots sometimes kill their sons, even though this leads to a maladaptive shortage of males in the next generation. In this species, females compete for nest holes that vary in quality, depending on how flood-prone they are. Females with better nests attract more males, who assist with chick feeding. It turns out that daughters are cheaper to raise, and fledge a week earlier than sons, so it makes short-term sense for single females with poorer quality nest holes to kill their sons in favor of giving their daughters a better chance. Daughters raised alone had a higher chance of surviving to adulthood than those raised with a brother. In contrast, females with less flood-prone nests had more male help, and could devote all their energy to defending the sought-after nest, while their mates fed the chicks. Infanticide of this nature is uncommon because most birds are impossible to sex as young chicks, and the older the chicks get, the less a parent saves by killing them. Eclectus parrots hatch with their very different sex-specific colors (males are green, females are red with blue) already printed on their down feathers making it easy for mothers to tell the sexes apart.

TOP LEFT

Tree swallows nest in holes, and a male that usurps a nest kills all the loser's chicks so that he can start a family as soon as possible with the female.

LEFT

Male eclectus parrots are green, while their Medea-like mates are red and blue.

PARENT–OFFSPRING CONFLICT

Raising children is essentially a cooperative venture, in that the ultimate evolutionary success for a parent's genes is to get as many copies of themselves as possible into surviving descendants.

BELOW

When a female mates with more than one male, this sets up a genetic conflict of interest between genes inherited from different parents (shown by colored dots in this hypothetical family). In the chick circled, the red gene from its mother has a 50% chance of being in each of its siblings. In contrast, the purple gene from its father has only a 25% chance of being in one of the other chicks, assuming that the two males have equal chances of fathering chicks in this nest. The less related the chicks in the brood are, the more conflict one would expect to see between siblings and parents.

The interests of parents and offspring should be aligned, but nature is rife with examples of parent–offspring conflict, such as weaning tantrums, in which children are demanding more than parents wish to give.

In sexually reproducing organisms like birds and mammals, each child only shares half its genes with each parent. Similarly, each child is 50 percent related to its full siblings, because a gene in one child has a 50 percent chance of being in its sibling. As most birds are not fully monogamous, siblings are often less related to each other than this. A parent would have the most offspring over the course of its lifetime by dividing all its resources evenly among them. At some point, it has to hold back from giving too much to any one child at the expense of the other children, including future offspring. This level of optimal parental care will always be less than the optimal amount a child has evolved to demand, because that child is less related to its siblings than to itself. The upshot of this asymmetry in relatedness is that we should see conflicts between parents and offspring over the length and amount of care.

Parent genes

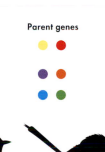

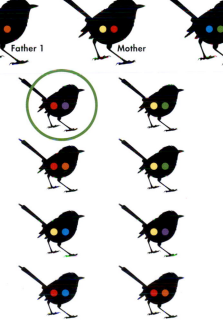

COMPETITION AND MANIPULATION

Bird siblings compete for finite parental resources, both within and between broods. They are also vehicles for the competing interests of their parents' genes. Most males are less likely to have as large a genetic stake in a brood as females, because of the high rate of cuckoldry.

As a result, the chicks in a brood with multiple fathers should be more selfishly demanding, at the expense of both their less related siblings and their unrelated fathers. As expected, the chicks in altricial species with higher cuckoldry rates have more vociferous begging calls. In addition, blue tit and zebra finch chicks are able to reduce the volume of their begging calls in response to the smell of familiar brood mates, but beg more loudly when exposed to the smell of an unfamiliar brood.

Species with precocial chicks are more likely to suffer from intraspecific brood parasitism, in which some females sneak their eggs into the nests of a different pair. Precocial chicks require less care, and are consequently less costly for a host parent to rear.

Of 97 rail species, 36 have ornamented chicks and larger broods with multiple parentage, both because of parasitism by extra-pair females and cuckoldry. As a result, chicks in a brood are less likely to share a genetic interest in each other or the parents raising them, which leads to increased competition and selfishness in the form of more exaggerated begging signals.

Brood parasites are the ultimate selfish beggars, because they are completely unrelated to other chicks in a nest, and to the adults rearing them, so there is nothing to gain by showing restraint and everything to gain by crying lustily. Cuckoo chicks manipulate their reed warbler foster parents into delivering more food by displaying an extra large gape, and by calling with more rapid trills to simulate hungrier reed warbler chicks.

Playing favorites

One way to minimize squabbling between one's offspring is to separate them. In birds, this practice is called brood divisioning, and it can take several forms. Among songbirds with altricial chicks, like robins and dunnocks, each parent preferentially feeds a subset of chicks, so that, rather than having to decide between six jostling mouths, each can concentrate on feeding three.

Brood divisioning is even easier for precocial species, because the chicks are mobile from an early age. In the first few weeks after hatching, coot chicks are fed by both parents, but each parent quickly becomes responsible for half the chicks, and within their half of the brood they each develop a favorite chick who gets the most food and attention.

LEFT

Blue tit chicks identify genetic relatives using smell, and beg more loudly and selfishly when exposed to the smell of an unfamiliar brood.

ABOVE

Grebes take brood divisioning a step further by developing favorites, who always know if they should climb up for a rest on their mother or father.

Crèches and chases

Parents that feed far out at sea, like penguins and pelicans, often leave their older chicks in crèches. When they return with food, they have to pick out their own offspring from the swarm of hungry chicks. Parents and offspring find each other with unerring accuracy in these situations, but three penguin species proceed to engage in an unusual behavior in which chicks chase their parents before being fed.

Rather than feeding their twins immediately, Adélie, chinstrap, and gentoo penguins turn around and run away, hotly pursued by their demanding chicks. The parent will pause briefly to feed a chick who has managed to catch up, and then takes off again as soon as the other approaches. This madcap feeding process is inefficient, and biologists now think its main function is brood divisioning to minimize competition between chicks. Supporting this interpretation is the fact that parents don't usually bother to make a single child chase them. They also tend to stop running the moment their chicks are separated enough for them to feed the leader. Lastly, attempts to feed two chicks at once result in wasted food, because no one retrieves food that gets dropped in the scrum.

Fair treatment

American coots often allow a third to half of all broods to die from neglect and starvation, and are a particularly good example of the effects of parental selection on the evolution of chick appeal. The younger coots in a large brood can suffer more from parental neglect, but they have one advantage—their brilliantly colored ornaments—that fade with age and are gone after three weeks. These ornaments attract the parents' attention, leading to higher feeding rates, particularly for the youngest chicks. By removing ornaments, biologists showed that parents fed brightly colored chicks more than black-headed chicks, but this only made a difference to the younger chicks, which were more dependent on parental provisioning. Even more telling, this parental preference was relative, not absolute, because the youngest were no longer the favorites in broods where all chicks had the same colored heads, but ornamented younger chicks were preferred over black older siblings.

In response to their irresistibly appealing youngest chicks, American coots also engage in a bizarre behavior to even out the playing field among their brood. Coot parents selectively punish the largest chicks by picking them up and shaking them vigorously in a behavior known as "tousling." Chicks suffer no lasting damage from this treatment, but respond like naughty children that have been spanked, and become temporarily subdued instead of clamoring for food and elbowing their smaller nest mates out of the way. Without the tousling, parents would probably end up with only one or two well-fed chicks instead of a larger brood.

TOP RIGHT

Gentoo penguin parents make their chicks chase them in an attempt to minimize sibling rivalry.

RIGHT

Like most rail species, American coots have highly ornamented, precocial chicks.

HONEST SIGNALS OF NEED

How do parents know when to respond to their children's cries for food? Costly signals offer a way out of an escalating arms race between parents and offspring, in which chicks are under ever-increasing selection to exaggerate their demands, while parents evolve to respond less.

In a situation analogous to the honesty inherent in expensive courtship displays, parents are more likely to believe begging that costs chicks something. Canary chicks that cry for longer gain weight more slowly, so screaming for food only pays when one is genuinely in need. Female canaries that anticipate a glut of food can increase the demands from their chicks by putting more testosterone into their eggs. When experimenters cross-fostered chicks to create a mismatch between the parental food supply and chick demand, fewer chicks survived, both in broods that demanded too much and those that demanded too little.

Males and females respond differently to different chick cues of need, and sometimes also use their partner as a source of information. Great tit males only respond to a visual display of the number of open mouths, whereas females respond to both the visual stimulus and vocal cries. However, if biologists enhance the amount of begging a female hears by playing recorded cries, both parents increase their food delivery rates. This is best explained by males using increased feeding by their mates as a cue to also deliver more food. Both parents benefit through rearing more chicks if they can share information about what chicks need.

LEFT

The carotenoids that chicks use to make their open mouths more appealing are the same precious pigments used as honest sexual signals of quality by adult birds. When biologists tried supplementing New Zealand hihi chicks with extra carotenoids, the chicks had redder gapes and got fed more.

INSURANCE CHICKS AND SIBLICIDE

Some species have an extra chick as insurance against hard times, and infanticide or siblicide culls the weak or unappealing.

Most birds of prey and some waterbirds like herons typically have more young than they can expect to raise, and, unlike many other birds, begin incubating the moment the first egg has been laid. This results in asynchronous hatching, where the first to hatch has a head start over everyone else, and so on until the last egg, which becomes the runt of the brood.

These runts only survive if the year happens to be a particularly good one with plenty of food, or if, by chance, one of the earlier eggs or chicks perishes. In this sense, the last chick in asynchronous brooders is reproductive insurance, and usually dies from neglect and starvation in more normal years because its older siblings outcompete it for food from the parents.

LEFT

Great blue herons typically produce an heir and a spare. These birds have obligate siblicide, in which one chick almost inevitably kills its sibling. The younger offspring typically survives only if the first egg was addled.

A barn owl feeds its
oldest and largest chick
a mouse.

Parents that rely on unpredictable food sources
can essentially use sibling conflicts to optimize
their own reproductive success. European
blackbirds feed on earthworms that only
emerge with rain, and their younger chicks
usually starve if rainfall is low. To see if
staggering the age and size of chicks was
really adaptive, biologists swapped chicks
around until all blackbird parents were raising
broods that were uniform in size. They found
that more chicks died in these broods, because
they were all so evenly matched that the
weakest took longer to die from starvation.

Fishing is a similarly unpredictable livelihood,
and many fish-eating species use sibling
competition to cull the weakest when times
are hard. Blue-footed boobies invariably have
twins, and when fish are scarce, and the older
chick is 20–25 percent below its expected
weight, it systematically pecks at its younger
sibling. The younger chick is eventually so
subdued it stops begging and starves to death.
Nazca boobies and brown pelicans have
obligate siblicide, in which the second egg
is pure insurance and the younger chick only
survives if its older sibling never hatched.

In addition to unpredictable food, large,
indivisible food also tends to increase sibling
aggression. Barn owls may have anywhere from
one to nine chicks, yet parents can only feed one
chick at a time, and deliver food about once an
hour. With chicks having to eat three to four times
a day, this creates large asymmetries in hunger
levels, and the youngest chick can be shoved out
of the nest during a feeding skirmish. Chicks will
sit on their hard-earned food to prevent their
siblings from stealing any.

In contrast to delivering relatively few large food
items like dead fish or game, parrot parents
regurgitate food for their chicks. This allows the
parents to be in more control, and to distribute
food more evenly among their brood. Siblicide is
much less common among parrots than among
other birds with asynchronous hatching. Siblicide
is also rare in glossy ibis, which feed the smallest
chicks preferentially.

A blue-footed booby chick solicits
food from a parent after it has
eliminated any competition from
its younger sibling.

SURVIVAL SKILLS

There are many things to learn before fledglings can graduate into fully functional, self-sufficient adults. Although many species are born with a few basic templates, it takes experience to hone their instinctive skills.

Leaving the nest is often the most dangerous time for young birds. Fledgling songbirds, which often spend some time out of the nest and on the ground before learning to fly, are easy prey.

Learning to communicate

Just as humans have evolved the hardware to learn languages, but need early exposure to develop this ability, young songbirds are born with specialized neural circuitry but require a tutor at just the right age to learn how to sing their species-specific song. Young white-crowned sparrows can be taught to sing different regional dialects by swapping their native dialect with a tape recorder playing the song of a male who lives in a different state. However, closed learners don't have the capacity to learn the songs of completely different species. Nor can they ever learn to sing properly if they miss hearing a tutor at the sensitive period during their development.

Open learners, including parrots, learn more than songs. Green-rumped parrotlets learn individual contact calls from their parents, and eventually develop signature calls as adults. Similarly, by cross-fostering yellow-naped Amazon chicks, biologists have found that family resemblances in these vocal signatures are purely the result of culture, not genetics, because they grow up sounding more like their foster parents and siblings than their biological family.

One hypothesis for why parrots, in particular, are so clever at learning a vast range of sounds and concepts is their very tricky diets. Parrots eat a lot of dangerous food such as seeds that are highly defended with poisons to discourage digestion, and this diet requires small amounts of a large range of things, as well as special knowledge of where to look for antidotes such as clay that binds toxins. Almost all this knowledge about food has to be learned, and parrot culture and communication has kept valuable and specialized knowledge alive.

RIGHT

Parrots, like the yellow-naped Amazon, are open learners, which incorporate new sounds into their repertoires throughout life.

Learning to feed

Most fledglings get a head start in life by graduating from the nest heavier than their parents, and many species continue to receive parental assistance and coaching even after they have fledged the nest.

Keas are highly social and rather brash parrots in the mountains of New Zealand. Unlike most parrots, who are highly suspicious of new things, keas are constantly experimenting and exploring, especially before they are fully grown. However, it isn't until keas reach reproductive age at about five that they can find food efficiently. Two-year-old juveniles have a special hunching, cowering posture that appeases adults enough for them to be shown food. In contrast, three- to four-year-old subadults graduate to mostly stealing their food from adults.

Secretary birds on the African plains specialize in rodents and snakes. Young birds learn how to hunt from their parents before they can competently stomp on their prey with their feet. In contrast, I once spent an entertaining afternoon by a mountain stream watching American dipper fledglings learning to feed through trial and error. Dippers are unique diving songbirds that eat crustaceans at the bottom of fast-flowing streams. Adults can propel themselves through the rapids with their powerful wingbeats and hop out on a rock within seconds, looking immaculate and plump, and holding a tasty snack. In contrast, the still fluffy fledglings I watched kept slipping and falling off their rocks into the rapids, and eventually emerged damp and bedraggled, with nothing more than a piece

FAR LEFT

An American dipper, or water ouzel, diving skillfully for food in a mountain stream. It takes practice for young dippers to reach the same level of expertise.

LEFT

Secretary birds are birds of prey that live on the grasslands of Africa. Young birds learn to use their feet to kill snakes, which are a popular prey item for these raptors.

FAR LEFT

Jacamars have a dangerous diet of butterflies that have evolved to mimic the patterns of highly toxic species. Young jacamars must learn which patterns to avoid. This in turn selects for better mimicry of the toxic butterflies by the tasty species, so as to avoid being gobbled up.

RIGHT

Redheads are ducks that know instinctively where to migrate without having to follow adults on the traditional route first.

of damp weed, which they would spit out in disgust. These failures happened repeatedly, and I can only imagine they would have spent the evening hungry had it not been for their parents, who were feeding nearby. The adults could occasionally be persuaded to part with their food if they happened to come up near one of their offspring, who would immediately switch from trying to be independent to frantically screaming for food.

Learning to migrate

Most birds that breed at higher latitudes migrate to warmer climates during the nonbreeding season. The challenge for a fledgling is knowing when and where to migrate. Young cranes and goslings follow their parents on their first migratory flights. For other species, migration routes as well as the impulse to migrate are largely under genetic control. Populations of blackcaps and dark-eyed juncos have evolved to stop migrating in places that are sufficiently comfortable in winter, making the bother of biannual travel unnecessary. Ducks seem to know instinctively where to migrate, because redheads often parasitize other ducks, but their fledglings still migrate to the Gulf of Mexico instead of following their canvasback host parents to their wintering grounds on the East Coast of North America. Similarly, adult pectoral sandpipers leave their chicks before they fledge, but the fledglings are able to migrate unaided.

FIVE

SEX ROLE REVERSALS

While most male birds have evolved to be the ornamented and aggressive sex that may leave females "holding the baby," other birds, like this male northern jacana, are the exceptions that prove the rule that an early asymmetry in parental investment leads to the evolution of showy males and choosy females, and sometimes vice versa.

A SPECTRUM OF SEX ROLES

In most mammals, males are larger, showier, more aggressive, and more ardent than females, who tend to be coy, choosy, and caring. However, this mammal-centered view of sex roles does not fit cleanly for many birds.

Reversed sex roles fall along a continuum, with species showing different combinations of reversed size, ornamentation, courtship behavior, mating systems, and parental investment and care. Australian brushturkeys are polygynous, with colorful males and drab females, yet the male is solely responsible for building and maintaining the nesting mound. Among birds of prey, females are the larger sex, and both sexes care for offspring equally. In the most extreme examples of sex-role-reversed birds, like jacanas and black coucals, females compete with each other for territories and a harem of males, who perform all parental care duties.

The paleognaths are a group of largely flightless birds that includes ostriches, kiwis, and extinct moas. Most paleognaths have paternal incubation, with a few species that are biparental.

LEFT

The extinct giant moas of New Zealand are among the largest birds to have evolved, and provide an intriguing example of sex-reversed body sizes. At 80–180 lbs (35–85 kg), male moas were considerably smaller than females, which could weigh up to a whopping 530 lbs (240 kg) —on a par with the heaviest sumo wrestlers. We don't know why female moas evolved to be so much larger than males, but based on living species, this size disparity could have been driven by the need to lay larger eggs and also by competition between females.

Backyard Bird

WOODPECKERS

It is relatively easy for woodpeckers to alter their breeding strategy in a given season, because males already perform at least an equal share of parental duties. Male Lewis's woodpeckers invest more than females in incubation and feeding chicks.

rather than just during the breeding season. Female streak-backed orioles are hard to distinguish from males, because many females are so bright. Biologists found female orioles attacked artificially augmented female decoys more vigorously. Males were less aggressive toward these more brilliant female intruders, whereas average colored decoys elicited similar levels of aggression from both sexes.

Flexible sex roles

Sex roles can be thought of as suites of strategies to maximize reproductive output. Some of these roles, such as fighting for mates and territories, or laying eggs, require physical adaptations that remain throughout an individual's lifetime, and tend to evolve as fixed differences between the sexes. In contrast, other roles, such as how much time and energy to devote to acquiring mates or caring for a brood of chicks, can often vary flexibly within the life of an individual bird.

BELOW

Female streak-backed orioles that are more colorful elicit more aggression from other females.

Female North Island brown kiwis lay the largest eggs of any bird in proportion to their own weight, but it is males who lose weight incubating the eggs for about two months. A single ostrich nest can contain about 20 eggs laid by different females. The eggs are incubated by a male and a dominant female, who often rolls the eggs of other females out, to make incubation more manageable. Greater rhea females also move between the nests of multiple males, who are incubating an average of almost 30 eggs.

Among socially monogamous birds with conventional sex roles, tropical females tend to be more colorful than their temperate counterparts. Signaling to members of the same sex is probably important for many tropical birds that occupy territories year-round,

WHY WOMEN'S LIB?

The bird behavior literature from the 1970s onward is full of terms like "emancipation from care," and some of the most popular hypotheses for reversed sex roles depend upon freedom from parental care. This could tell us just as much about biologists as it does about birds, but while many reproductive roles, from courtship to care, tend to come in a package, we still lack a unifying explanation for sex role reversals.

This unconvincing consensus is partially due to a lack of data. In only 5 percent of birds, females compete for mates while males perform more care. A mere 1 percent have exclusively male care, and the vast majority of these species have precocial young, which don't need much care at all.

Four eggs good, more eggs better

Physiology and life history predispose some species to adopt reversed sex roles. Precocial offspring and smaller clutches make it possible for a single parent to raise just as many offspring as two or more adults. Sex role reversals are especially common among shorebirds, which fit both of these requirements. All shorebirds are physiologically constrained to lay a maximum of four large eggs per clutch. This is a handy number to fit under a single incubating adult. Any more, and some eggs would be left out in the cold. Shorebird chicks are also mobile and well insulated from the moment of hatching, making it easier for mothers to abscond.

Migration and polyandry

Among shorebirds, there is a strong positive correlation between migration distance and polyandry. However, no one knows if emancipation from care allowed more females to leave earlier to migrate, or if species with longer migration distances evolved paternal care to allow females to refuel after the expensive act of egg laying. Female willets and common sandpipers frequently abandon their families to migrate before the chicks are fledged, while long-billed curlew and spotted redshank females often leave before their eggs have even hatched. Having so much male help makes it but a small step from social monogamy to polyandry, and female Kentish plovers, dotterel, and dunlin frequently leave when their chicks are a few days old to begin a new family with a different male.

TOP RIGHT
Female long-billed curlews are shorebirds that usually abandon their brood 2–3 weeks after the young hatch, leaving the male to perform all the parental care solo.

RIGHT
Like many other shorebirds, female Eurasian dotterel merely lay their eggs and leave, while their less colorful mates remain to incubate the eggs and care for the chicks.

HOUSING AND MATE AVAILABILITY

In species with male care, territory quality and remating opportunities are major factors that influence the degree of sex role reversal. Given the fairly fixed conditions of reproduction and development that characterize groups of birds, species and individuals can adopt sex roles that fit a particular ecological and social environment.

Mating markets are driven by nesting densities as well as adult sex ratios, both of which influence sex differences in the reproductive benefits of deserting a current mate. Biologists have found that nesting density is the only consistent difference between bird families in which all species have single fathers or single mothers. Birds with male-only care breed farther apart, usually with less than one nest every 10 hectares, than those with female-only care. At low densities, females are more likely to benefit from deserting first because they can rebreed more successfully than males. This is because females can store sperm, and don't need to find another mate to start a second family, while deserting males are less likely to find a female ready for mating.

Whichever sex is in shorter supply has a greater choice of partners and can demand more from the partners they choose. For instance, female Kentish plovers are more likely to mate with multiple males in populations where males outnumber females. This is because a female that deserts her mate can find a new partner almost immediately, whereas males that try to leave their families have to look for a mate for much longer, losing valuable breeding time. Similarly, whichever sex is first to desert in Florida snail kite society depends largely on who can remate more quickly, so the mating system in a particular population depends on the sex ratio.

TOP LEFT

Snail kites are flexibly polyandrous when females are the rarer sex, and therefore quicker to remate.

LEFT

Female Kentish plovers frequently leave their chicks entirely in the care of the males.

Backyard Bird

WILLOW TITS

Willow tits in Finland are socially monogamous, but partially role reversed, in that females are more territorial than males early in the breeding season. This is because females perform all incubation, making renesting more costly than for males. The roles switch later in a nesting cycle, when males take over the job of guarding the territory. This is because the breeding season is so short that males have almost no chance of starting again once the chicks are growing, so they have little incentive to leave and attempt to gain a second family.

Sex role reversals often reach extremes when there is a seasonal surfeit of food. Phalaropes breed near water at high latitudes, where their insect food, such as emerging midges and mosquitoes, comes in big, brief, unpredictable bursts. Since a female red-necked phalarope can't lay more than four eggs per clutch, the best way for her to capitalize on a brief glut of midges is to lay a quick succession of clutches for a series of mates. A male's best response is to literally sit tight and tend his nest alone, because the vast quantity of food creates a situation where females are limited by how many mates they can find, rather than the number of offspring they can produce. This state of affairs is self-reinforcing, because when males are all tied up taking care of chicks, there are fewer available for females to mate with. Males are in short supply, so larger, more competitive, showier females have an advantage. It also pays females to concentrate all their time and energy on acquiring mates, rather than on parental care.

BELOW LEFT AND BELOW

Red-necked phalaropes are polyandrous shorebirds that breed on the Arctic tundra where the season is short. Females (left) are larger and more colorful than males (right).

Safety and sustenance

In contrast to the sequential polyandry practiced by phalaropes, other role-reversed shorebird females can afford to capitalize on food-rich territories by commanding enough property to support multiple males at once. Simultaneous polyandry is also a good strategy in dangerous neighborhoods, because if nests fail often, the well-fed females can quickly lay a replacement clutch for whichever male lost his offspring to a predator. Both sexes maximize reproduction by having males perform all or most of the parental care, freeing females to become egg-laying machines. A female spotted sandpiper lays five clutches in eight weeks, which is the equivalent of four times her own body weight in eggs. Spotted sandpipers rely on bursts of insect food hatching from freshwater ponds and lakes, and their nests—scrapes on the ground within 100 yards (90 meters) of the water—are vulnerable to many predators.

BELOW

Spotted sandpiper males are always single parents because females mate with and lay eggs for multiple males in quick succession.

AVIAN AMAZONS

Stiff competition for territories and mates has driven the evolution of larger and louder females among birds with reversed sex roles. Among such species, the black coucal is the most extreme.

ABOVE

The only altricial bird with sex role reversal, black coucals suffer high losses to predation, and males are more limited by how soon their mates can lay a replacement clutch than by the costs of parental care.

Female black coucals are dominant over males, whom they outweigh by 70 percent, and sing to defend territories and attract multiple mates. Males are not loud or territorial, and perform all childcare duties, from nesting to feeding chicks.

Singing for one's status

Biologists have found that larger black coucal females have deeper voices, and that females use this information when communicating with competitors. When biologists played recordings of a lower-pitched song, they found that listening females hesitated for longer before replying or approaching, kept their distance, and spent less time near the loudspeaker. When they did pluck up the courage to respond, they would deepen their own voices, as if to artificially inflate their own size in the ears of a competitor.

LEFT

Wattled jacanas are South American shorebirds. Like all jacana species, they have long toes that enable them to walk on lily pads, earning them names like "lily trotter" or "Jesus bird." Females are larger than males, and polyandrous.

Size matters

When females are under the dual pressures of fighting to hold territories and pumping out a lot of eggs, they are under strong selection to be big. Female cassowaries are larger, more brightly colored, and more aggressive in territorial disputes than males, whereas the smaller, drabber males perform all parental care duties solo.

Among shorebirds, northern and wattled jacanas of the Americas show the most extreme reversed sex roles. Females fight each other to control a limited supply of breeding territories and are completely dominant over males, which are about half their size. Each female controls the territories of up to four males, and if she manages to take over a neighboring female's mate, she kills his chicks so that, with no offspring to raise, he is available for remating sooner. The larger a female's territory, the more males she can accommodate.

Male bronze-winged jacanas from India have a different agenda, which makes it a bit harder for a female to acquire a large harem of males. Larger males defend larger territories, so females with larger harems have more small mates squeezed together. Like their American relatives, bronze-winged jacanas are thoroughly role reversed. When biologists removed individuals to see how the sexes establish new territories, they found that, unlike birds with conventional sex roles, male jacanas often stay put when their mate disappears, whereas females quickly rearrange their territories to take over as many new mates as possible.

LEFT

The southern cassowary is one of three cassowary species from Australasia. These ratite females are larger than their male counterparts, and defend territories that encompass those of multiple males.

BEAUTY CONTESTS

Male birds don't have a monopoly on beauty, even if they usually have to work harder to attract females than the other way around. Among many socially monogamous birds, females are ornamented because males also prefer beautiful mates.

Male crested auklets are more likely to approach and display to females with longer head crests. Similarly, female barn swallows with longer tails mate earlier, and bluethroat and blue tit females with brighter ornaments get more attention from males.

As with male ornaments, beauty can be a signal of quality. Female barn swallows with longer tails laid more eggs and raised more chicks, even though they didn't actually care more for the chicks than less attractive females. Similarly, female bluethroats with brighter throat patches were also larger, but were not the most conscientious mothers.

Biologists are starting to recognize an increasing number of ornamented females among socially monogamous birds, but, as with extreme size disparity, the flashiest females are found among sex-role-reversed species with choosy males.

Beauty spots

Female spotted sandpipers compete with each other to retain the most productive territories and attract as many males as possible. This reversed sexual selection has led to females outweighing their mates by 25 percent.

Reversed sexual selection also explains why female spotted sandpipers have bigger spots, the better to attract males with. Females are more spotty than males, and neither sex invests in spots during the nonbreeding season, suggesting they carry a cost. In addition, individuals can change their spots across seasons, allowing them to flexibly signal their condition. Females with larger spots were bigger and had fewer mites. Similarly, those covered in a larger proportion of dark spots had a higher red blood cell count, suggesting they were in better health. The melanin pigments that make up spots are known to help protect feathers from abrasion and damage by bacteria and mites. There was a much weaker relationship between spottiness and condition among males.

BELOW

Both female and male crested auklets assess potential mates on the basis of their crest length and strong tangerine scent.

STAY-AT-HOME DADS

Childcare is an expensive business, and not one to be taken for granted. The sex that performs more care has less time to remate, and tends to be the sex in higher demand. When sex roles are reversed, males are limited by the number of offspring they can afford to rear, whereas females are limited by the number of matings they can win.

A male wattled jacana takes three months to raise a brood to maturity, which puts him in a cruel bind, because as the sole caretaker he cannot abandon his current reproductive investment without sacrificing it completely.

Evolution and male-only care

Classic examples of male-only care are found in the ratites, tinamous, kiwis, and shorebirds, all of which have precocial, self-sufficient chicks. Male-only care has evolved only once among altricial birds, in the African black coucal, which is a species of nonparasitic cuckoo.

LEFT

The plains wanderer of New South Wales is so different from other bird species that it has been placed in a taxonomic family of its own. One of the few things we know about them is that males are solo parents.

Sex Role Reversals

Hormonal sensitivity and sex roles

Biologists have long been intrigued by how males in role-reversed species manage to suppress their "male" urges and stick to so much stolid caregiving. Buttonquail are dumpy terrestrial birds that look so much like true quail that they used to be placed in the chicken and pheasant family. Evidence from fossils, DNA, and anatomy now place these 16–17 species within the large group of shorebirds. Their sex-role-reversed behavior should come as less of a surprise, since shorebirds are the avian poster children for sex role diversity. Male buttonquail perform all parental care, whereas the larger,

more colorful females court them with gifts of food and booming calls produced by a specially adapted throat organ. Rather like true quail, barred buttonquail are common and easily raised in captivity, so they are perfect for studying the hormonal underpinnings of sex role reversals.

Unfortunately, as with most of science, answers are not as straightforward as one might hope. Female buttonquail do not have higher testosterone levels than males. Similarly, Wilson's phalaropes did not have different hormone profiles from other species with conventional sex roles. As the only completely role-reversed species with altricial young, black coucal males have a lot more work to do rearing chicks than the fathers of precocial chicks. In spite of having a completely atrophied left testis, male black coucals do not have lower levels of circulating testosterone than females.

However, when researchers looked at the brains of female buttonquail and black coucals, they found that female brains had more androgen receptors than male brains. It is not circulating levels of sex hormone that alters sex-specific behavior, but the sensitivity to these hormones in specific parts of the brain.

LEFT

The female barred buttonquail is larger, more aggressive, and less caring than her mates. Her brain is more sensitive to androgens such as testosterone.

MALES IN BONDAGE

Most birds with purely paternal care have precocial offspring that are independent within days of hatching. However, males still pay a significant cost as the primary or sole caregivers.

A male emu can lose a fifth of his body weight after incubating the eggs solo. To add insult to injury, not all the eggs may be his, because females of many role-reversed species mate with multiple males. What's to keep a male faithful if he has no genetic stake in the offspring he is caring for?

Biologists used to think that polyandrous species sustained paternal care on the strength of a male's certainty of paternity. In other words, sex-role-reversed species should have lower rates of extra-pair paternity, as an incentive to keep a male at his job of rearing chicks. However, the evidence that certainty of paternity is necessary for paternal care is decidedly mixed. Tinamou males are sole incubators, but commonly incubate the eggs of other males, because females mate freely with multiple males.

In the case of Wilson's phalaropes, biologists have never recorded extra-pair paternity. This is probably because males can afford to desert and renest if they suspect females of polyandry, and females find it harder to remate later in the breeding season because all the mature males are tied up with families.

TOP RIGHT

Like most ratites, emu males perform the majority of parental care, regardless of how many of the chicks are genetically theirs.

RIGHT

Wilson's phalaropes are highly polyandrous, and the larger, more colorful females (shown) leave parental care to males.

In contrast, sequentially polyandrous red phalarope and spotted sandpiper females store sperm from previous matings, so that any male that is not a first mate is almost bound to be cuckolded by the male that came before him. Even if a male red phalarope wanted to abandon a current brood when he suspected adultery, he would be hard put to renest and rear any chicks successfully because of the short breeding season in the Arctic. Nevertheless, EPP rates in these role-reversed shorebirds are less than 10 percent, which is less than in most socially monogamous species.

Jacana males in simultaneously polyandrous relationships have to accept a 40 percent chance of EPP, and an average of almost a fifth of their brood will not be fathered by them. This is because a wattled or bronze-winged jacana female copulates with all her mates and lays

ABOVE

This male African jacana is guarding a nest with eggs that he will have to incubate solo because his mate is busy laying eggs for other males.

clutches of eggs for each male in turn, not just the one whose nest she is laying in at the moment. Male jacanas in polyandrous relationships are not cuckolded by stored sperm from a previous male, but from the female mating with all her mates in quick succession. A male jacana with a powerful mate has really drawn the short end of the reproductive stick, because as a member of a larger harem, he has to wait even longer for her to confer a clutch of eggs on him, and the longer he waits, the lower his chances of fathering the eggs he has to care for.

Attempts to enforce female fidelity

A female jacana or spotted sandpiper controls a territory that encompasses the jealously guarded individual fiefdoms of her mates. This makes it much harder for males to enforce any kind of mating fidelity, because none of them would allow each other into their tiny territories in the first place, and because the females are larger and would easily overpower any petulant male.

In an attempt to keep females from visiting their other mates, bronze-winged jacana males call for attention. If a male pheasant-tailed jacana has no mating visits from his mate, he tosses out any eggs she lays for him, because they can't possibly have been fathered by him.

In nonterritorial, sequentially polyandrous shorebirds like Wilson's and red phalaropes, males attempt to capitalize on the fact that females often store enough sperm to fertilize subsequent clutches cared for by other males. Males try to be the first a female mates with. Failing that, they become increasingly ardent toward the end of egg laying, when their mate no longer needs sperm to fertilize the current clutch, and is more likely to store enough sperm to cuckold her next mate. Dotterel males adopt the closest thing a subordinate male has to mate guarding, and play hard to get until right before the female is about to start egg laying. These delaying tactics reduce the chances of her fertilizing their clutch with sperm from a previous male.

BELOW

Like other jacana species, male pheasant-tailed jacanas of India perform all parental care. They ensure the eggs they incubate are genetically theirs by rejecting any eggs laid by a female that they haven't just mated with.

COOPERATIVE POLYANDRY

This chapter has focused on polyandrous systems where each male has his own nest and clutch of eggs to tend. In cooperative polyandry, males don't just share a female; they also share a communal nest and care for the eggs and chicks as a group.

ABOVE

Kagu (shown courting) are only found on the island of New Caledonia, and are an avian example of cooperative polyandry among male kin. Male kagu form clans that cooperate to defend a communal territory. A clan comprises up to five families, each with a breeding female who is shared by up to three brothers. All the breeding males in a clan are relatives, whereas all breeding females are unrelated. Adult sons in the clan will often visit their parents, even after establishing their own polyandrous families with an immigrant female.

Suitable nest holes are highly sought after by eclectus parrots of Melanesia. Females are the more colorful, dominant sex that competes for these precious holes, which they proceed to guard by remaining in the hole, attended by their harem of males. Male eclectus parrots share the work of feeding their mate and chicks, and are a well-camouflaged green, in contrast to the brilliant blue and red females. Holes are in such short supply that as many as seven males are forced to share a female if they want to breed at all.

In contrast to classical polyandry, which often coincides with very rich food sources, cooperative polyandry results from severe scarcity and often intense competition for those precious resources. A group of Galapagos hawks can guard a territory for more than 10 years, and an adult's annual survival on a territory is greater than 90 percent. In contrast, food and territories are so hard to come by that offspring survival is very low. At best, a group can expect to raise one fledgling in a year. Males team up because larger groups are better at finding food and defending a territory.

Female Galapagos hawks have an average of five mates who share paternity and paternal care with very little evidence of infighting or shirking. The males are usually unrelated to each other, but they will contribute equally to chick care regardless of how many times they mated. Unlike the socially monogamous birds of previous chapters, these males do not offer care depending on their odds of fathering the chicks. Biologists have observed very little aggression between them, no dominance hierarchy, and no attempts to mate guard. In contrast, Harris's hawks of Arizona have a strong dominance hierarchy, so although a group of males cooperates to hunt, the alpha males, who father most of the chicks, do most of the parental care.

When males are the main incubators, the resulting breeding system depends on how easily a male can incubate a communal clutch. Cooperative polyandry, common among rails like the pukeko, usually results when more eggs from more females would compromise hatching success, and more than one male increases offspring success. In contrast, if offspring survive just as well with one as with two parents, serial polyandry results, as seen among many sandpipers, phalaropes, and buttonquail.

ABOVE

These Galapagos hawks cooperate to care for chicks even if they haven't fathered them.

LEFT

Cooperatively polyandrous pukeko males tend to be unrelated to each other, and share parental care duties equally. This is surprising, because there is a strong hierarchy within a pukeko harem of unrelated males, and the dominant male sires most of the young.

SIX

GROUP BREEDING

Sociable weavers nest in huge bird condominiums, and rooks commonly nest in colonies, but some species take group nesting a step further and care for chicks cooperatively. A tenth of bird species, including the white-fronted bee-eaters of Africa (left), have helpers at the nest who forgo their own reproductive opportunities to assist the breeding pair.

COOPERATIVE BREEDING

Human parents often receive a lot of help from others in raising their children, and so do about one-tenth of birds. Depending on the species or even the population, there are many reasons why cooperative breeding has evolved.

Evolutionary origins

Cooperative breeding, in which more than two adults tend a single nest, occurs on every continent other than Antarctica. As early as 1841, the eminent British ornithologist John Gould described the behavior in fairywrens, which have continued to play a major part in advancing our understanding of cooperative breeding. In 1935, an American biologist coined the phrase "helpers at the nest" to describe the breeding behavior of individually marked bushtits, brown jays, and cactus wrens.

ABOVE LEFT

Red-cockaded woodpeckers are an endangered species in the American southeast, due to the loss of the longleaf pines on which they specialize. These cooperative breeders rely on large groups to defend precious nesting trees.

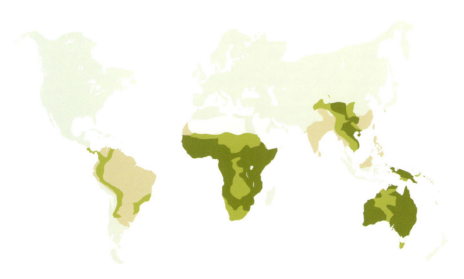

LEFT

Cooperative breeders are more common in the tropics, and show a striking overlap in distribution with obligate brood parasites (see page 171). The darker the color, the greater the proportion of bird species that breed cooperatively.

In an analysis of more than 3,000 bird species, biologists found that cooperative breeding almost always evolves from an intermediate step of family living. Of the species they studied, 31 percent have young adults who remain home with their parents but don't actually help with the raising of younger siblings. 13 percent live in family groups, and relatives assist with care. Just 1 percent of species breed cooperatively, but without related helpers.

This analysis also explains a conundrum that has stumped students of cooperative breeding for decades. Many cooperative breeders, such as red-cockaded woodpeckers and superb fairywrens, are found in stable and productive environments. Saturated property markets and the benefits of having a group to defend a coveted resource are the primary explanations for why these young birds stay home and help their elders rather than starting as independent breeders.

Conversely, a disproportionately high number of cooperative breeders, including white-winged choughs, superb starlings, and southern pied babblers, live in harsh and unpredictable parts of Australia and Africa. Here, the "hard life hypothesis" prevails as an explanation for group breeding: when the collective benefits are so great that they outweigh going it alone, individuals will invariably breed in groups as protection from unpredictable periods of starvation. Some groups will even kidnap young if they are unable to rear enough helpers of their own.

BELOW

The superb starling always breeds in large groups, regardless of conditions. The idea is that individuals experience such a high risk of failing to breed with the highly unpredictable rainfall patterns in central Kenya, that they always do best hedging their bets by breeding in groups. A stable reservoir of subordinates acts as a buffer against uncertainty.

The question of why both stable, rich environments and unstable, poor ones explain cooperative breeding is neatly solved by showing when and where bird lineages evolved family living and helping. The pattern is one in which young delay independent life in times of plenty, and, if conditions subsequently deteriorate, are driven to start helping rather than simply freeloading off their parents. This also explains why so many species live in families without helping, and why a disproportionately high number of cooperative breeders live in Australia, southern Africa, and northern South America, where the climate has become much harsher since past geological epochs.

Hogging the benefits

The benefits of staying are especially evident in species that delay dispersal but don't help. Siberian and Canada jays delay dispersal when intact spruce forest is limited and there are no territories for young couples to settle in. The strongest fledgling boots its subordinate siblings off the territory with aggressive "bill clicks," and becomes the sole beneficiary of parental nepotism.

TOP LEFT

White-winged choughs live in the harsh Australian bush, and are examples of the "hard life hypothesis" for cooperative breeding.

LEFT

Immigrant Siberian jay helpers are so harried and hungry that they develop more risk-prone personalities than their cosseted counterparts, and stop bothering to scan for danger before feeding.

Backyard Bird

CARRION CROWS

Carrion crows in parts of Spain and Switzerland breed cooperatively in groups of up to nine adults. Only one female breeds, but in addition to her social mate, the dominant male, she also mates with immigrant males, who are often his cousins. None of the adult offspring who remain home for up to four years breed, but all adults assist with defending a year-round territory, nest building, chick feeding, and removing fecal sacs from chicks in the nest.

Carrion crow parents are less tolerant of immigrants than of their grown offspring, whom they actively aid and protect. Dominant males keep immigrant male cobreeders away so their adult sons and daughters can eat before their superiors in the group hierarchy.

Breeding pairs share food with their adult offspring, and watch for danger so the latter can spend more time feeding. They also warn each other with alarm calls that signal the type of predator and the threat level. As a result, these delayed dispersers live longer and have more offspring over their lifetimes than their siblings, which had to leave home earlier and join an unrelated group, where they are treated as second-class citizens. However, all nepotism stops the moment one parent dies and is replaced. Lingerers leave within a month of a stepparent's arrival.

A leg up in life

Florida scrub jays breed in strictly monogamous pairs with help from their adult offspring, particularly their sons. One way for young adult males to start their own families is to bud off a portion of their parents' substantial territory. Settling on the family estate is not for the faint of heart, however. In their first three years, budding males are just as likely to fail as their brothers who disperse further. It is only those who persist beyond the third year that end up with a higher lifetime fitness.

Some breeding pairs chase away females attempting to pair with their adult sons, but many actively expand their territory as he is reaching pairing age, as if to provide him with a place to settle. Pairs that control larger tracts of high-quality land can increase their reproductive success not just by breeding more, but also by maximizing the chances of their sons producing grandchildren.

MARKET SATURATION

When all the homes in the best neighborhoods are taken, property owners can capitalize on the presence of other adults by accepting their help with maintenance, defense, and childcare.

Mexican jays are plural cooperative breeders, with multiple pairs in a group breeding together on a group territory. Why do their close relatives, the Florida scrub jays, invest in acquiring and defending just as large a territory when only one couple breeds, while the other adults must be content to remain as nonbreeding helpers? The answer lies in the highly specialized and limited post-fire scrub that Florida scrub jays require.

Florida scrub jay groups defend large territories all year round as a triple insurance policy. Firstly, fire keeps competition and predation low, and

larger territories are more likely to include burned areas. Secondly, the scrub required for the preferred scrub jay diet of particular acorns and arthropods requires frequent fires, but plots that have just been burned take a few years to recover. A large territory is more likely to encompass both current and future feeding sites. Thirdly, wealth breeds wealth, because fledglings reared on larger territories are heavier, which improves their odds of becoming breeders.

LEFT

Green woodhoopoes in Kenya nest in groups where tree cavities are scarce, but in pairs where cavities are plentiful.

ABOVE

Florida scrub jays reared on peanuts in the Florida suburbs are less likely to live, and grow up to be smaller than their country counterparts, leading to lower rates of survival and reproduction as adults.

Parents on larger territories also fledge more young that turn into nonbreeding helpers, and larger groups control larger territories in a self-perpetuating cycle. Large territories are so important for these birds that even pairs without helpers will devote extra effort into expanding and defending their territories, as if banking resources for future helpers.

Prize property

Red-cockaded woodpeckers differ from other woodpeckers in requiring live pine trees in which to drill their nest holes. The resin from the wounded tree serves as an excellent antipredator defense, as any snakes attempting to wriggle into the nest hole get stuck.

These woodpeckers from the southeastern US are endangered because the longleaf pines that they prize as nest trees require regular fire, so decades of fire suppression by humans has led to habitat loss. As suitable nest trees are so limited, territories tend to be large and stable, fiercely guarded by a monogamous breeding pair and up to five of their adult sons or the odd unrelated immigrant. Just by drilling more nesting holes, biologists induced several of these helpers to leave and set up breeding territories of their own, showing that helpers were biding their time as nonbreeders because of a saturated housing market. As an unexpected consequence of this academic experiment, conservationists refocused their efforts from reducing predation and increasing the size of protected areas to improving habitat quality through prescribed burns.

Adapting to circumstances

Seychelles warblers are one example of how some cooperative breeders are making the best of a bad job in a saturated housing market. These birds almost went extinct because of introduced rats and cats, and were reduced to a single cooperatively

ABOVE

Seychelles warblers are a prime example of helpers at the nest who stay at home because they have nowhere to set up a territory of their own.

breeding population on Cousin Island, which has been studied since 1981. Thanks to conservation efforts, the warblers soon filled all the available space on that island, and are now thriving on five islands in the Seychelles. Importantly, warblers transferred onto a new island stopped breeding in groups until they had saturated all the suitable habitat once again. When all the territories were occupied, groups could command higher-quality territories than pairs, improving the survival and lifetime reproductive success of group members.

FAMILY MATTERS

Any behavior that helps others will not evolve and persist unless it also propagates copies of the genes that programmed it in the first place. So most of the seemingly altruistic acts we see among cooperative breeders are really quite selfish when viewed from the perspective of the genes that are being indirectly propagated through the reproduction of relatives.

Voice recognition

Some birds have ways to identify kin so that they can target their help to where it is most likely to benefit their chances of genetic posterity. Western bluebirds can switch flexibly between breeding and helping over their lifetimes. Male bluebirds remain in touch with their fathers or even grandfathers after migrating away from their breeding grounds for the winter, and often return to settle near each other. Even male relatives that settle farther apart fly across other territories to help each other. Biologists recorded a male who became his father's helper, and when that nest failed too, both flew to help his grandfather. Since physical neighbors sound more similar, biologists think bluebirds learn and then remember their relatives' voices for the rest of their lives, creating a sort of mental vocal and spatial map so that they can direct their help where it matters most to their genes.

LEFT

An overwhelming majority of cooperative breeders help and accept help from genetic relatives. This is particularly convenient for white-fronted bee-eaters, which nest close together.

RIGHT

Western bluebird brothers settle close to their parents. The upshot of this limited dispersal is a kin neighborhood, where just by moving next door, one is likely to be helping a relative.

BELOW RIGHT

Among long-tailed tits, males disperse less than females, so most helpers at the nest tend to be male relatives.

Family signatures

Long-tailed tits are found all across Eurasia, and form wintering flocks that roost together for warmth. In spring, the flocks dissolve into breeding pairs with overlapping territories, but in populations with short breeding seasons and high predation rates, failed breeders turn to helping other flock members feed chicks. After DNA testing lots of birds, biologists found that over three-quarters of helpers assist first- or second-order relatives, and bring more food to chicks who are closer relatives. Curious as to whether individuals could distinguish between related and unrelated flock mates, they gave birds the choice between helping two neighbors nesting within a similar distance. Out of 17 birds, 16 chose to help their relatives over unrelated flock mates.

The next question was how long-tailed tits managed to recognize relatives. To test this, biologists cross-fostered half the chicks from several nests, and recorded everyone's calls. They found that unrelated birds raised together sounded more like their foster families than their genetic relatives, suggesting that long-tailed tits learn a family-specific signature from their carers. This system allows for relatives who have never met to recognize and help each other.

Choosing relatives

Although species such as long-tailed tits are only cooperative breeders when their own nests fail, species that always breed cooperatively also direct their help toward close relatives. Bell miners, chestnut-crowned babblers, and superb starlings all live in complex two-tiered societies, with large, social foraging groups that comprise smaller breeding groups with helpers. Males generally disperse less than females, making most of these societies patrilineal, so males stand to gain more from mutual help than females.

Bell miners mew when they attend a nest, and use this call to coordinate feeding visits as well as to identify relatives. The more closely related helpers are to the breeding male, the more food they bring. Chestnut-crowned babbler helpers are also mostly male, and helpers always feed the most closely related brood in the larger social group. Furthermore, individuals helping to rear full or half siblings worked three times as hard as those feeding less related chicks.

BELOW

Bell miners live in complex, two-tiered societies in Australia. Helpers tend to work harder to feed chicks that they are more closely related to.

DYNASTY DYNAMICS

Business and political empires often pass top positions to a male from within the family. At times, the line of succession is broken by an unrelated usurper, who must prove himself before gaining the position of power. Males in cooperative breeding bird societies attain top breeding status in much the same way.

Heirs to the throne

Bell miner and superb fairywren sons bide their time as nonbreeding helpers before eventually inheriting the dominant breeding position on their fathers' territories. Superb fairywren males can reign for as long as 10 years, which means their sons can have a long wait for the throne. Nevertheless, 60 percent of males inherit the top breeding position on the territory where they were born, even if it means being socially paired with their mother, the dominant female. Instead of waiting, other sons may form an alliance with an immigrant female and attempt to split the group territory, whereas others attempt

to pair with neighboring widows. Breeding female fairywrens are always in short supply because more females than males die during the breeding season.

Enemies from within

Immigrant males are very different from sons who stayed, in that they are generally unrelated to the rest of the breeding group. These unrelated males pose a threat to the breeding male and fail to gain indirectly by helping to raise kin. In many species, including merlins, hoopoes, and Puerto Rican todies, unpaired males join a group in the hope of sneaking a mating while ostensibly "helping" the breeding pair.

LEFT

Although some merlin "helpers at the nest" have been observed copulating with the breeding female, DNA fingerprinting finds no evidence that these extra-pair copulations are successful.

Curiously, some of these hopeful breeders, such as rufous vangas, riflemen, and white-browed scrubwrens, work just as hard, if not harder, than the breeding pair's genetic sons or the breeding male himself. Bell miner and pied kingfisher males provide a possible explanation for this apparently selfless behavior. In both species, a hardworking helper stands a better chance of rising to the position of breeding male when the reigning male dies. Some bell miner males have even been so assiduous with their attentions that breeding females were persuaded to leave their mates in favor of the helper that delivered the most food.

ABOVE

Pied kingfisher males work hard in the hope of rising to the position of breeding male.

RIGHT

A male rufous vanga at a nest in a dry forest area of Madagascar. The breeding male in a group may find himself challenged by unrelated males whose help outweighs his own work at the nest.

150

Collective benefits

There are other duties, besides feeding chicks, that often require a minimum group size, particularly to survive harsh environments. Superb starlings and stripe-backed wrens need a minimum group size to deter nest predators. Chestnut-crowned babblers and green woodhoopoes nesting in places that get very cold at night need enough warm bodies with which to huddle while roosting.

Southern pied babblers from the Kalahari Desert require all these services and more, including help with incubation, brooding, provisioning, teaching, escorting young, predator mobbing, territory defense, and sentinel duty. Of particular importance is their sophisticated sentinel system. Pied babblers face extremely high predation, and time spent watching for danger means less time for feeding. Whoever is on duty sings a "watchman's song," so the rest of the group knows they can forage in relative security.

In this species, helpers are especially useful to the breeding pair in drought years, when extra sentinels, soldiers, and babysitters are crucial. Chicks attended by more helpers experience a longer period of care, which improves their chances of becoming breeders when they grow up. Larger groups occupy larger territories, and can afford to divide the labor so that some helpers are defending the nest while others are foraging to feed chicks. Indeed, group living is so crucial that small groups that fail to breed (one of the best ways to recruit helpers) tend to go extinct.

RIGHT

Southern pied babblers struggle with temperatures above 95°F (35°C), and have to sit about dissipating heat instead of foraging. Climate change could make things even more untenable for small groups.

Kidnapping

There are two things small groups can do when desperate. One is to merge groups, which is highly unusual for such a territorial species as southern pied babblers, and results in an unstable and temporary alliance that dissolves the moment a drought is over. The other is to kidnap young from a neighboring group, and rear them as one of your own.

The kidnappers do this by initiating a border dispute. When the neighbors are distracted, one member of the smaller group sneaks into the heart of their neighbors' territory to look for chicks, hoping to elude any babysitters left behind to guard them. While holding bait in its bill, the kidnapper gives out soft food calls, a combination which the greedy youngsters find irresistible, and lures its unsuspecting victim safely across the border. At that point, the rest of the group abandon the decoy border dispute and join in tempting the kidnapped youngster into the heart of their own territory, where it is treated like one of their own and grows to become a helper for the kidnappers.

THE COSTS AND BENEFITS OF CARE

Helping can be costly to both sides. Subordinate Seychelles warblers that act as nannies end the breeding season in poorer condition than group members that didn't help. Breeders also face added competition for resources such as food, housing, and even mates, so helpers can be a hindrance. What determines who helps, why they help, and how helpful subordinates really are?

Public goods

Everyone benefits from public goods, which means that both breeders and subordinates care, even if some may care more than others. Arabian babblers roost in families, so parents tend to self-sacrificially take the ends of the linear huddle. Acorn woodpeckers huddle in communal roosts, and members of larger groups pay a lower individual heating cost, which improves survival.

Paying to stay

Some territorial breeders require "rent" payments in the form of childcare. In superb fairywrens, helpers "pay to stay," as breeding males punish helpers who shirk their duties. Biologists removed a helper for a day while there were chicks to feed. When they returned the helper, the breeding male attacked him so viciously and unrelentingly that the experiment had to be aborted.

Biding time

The home territory may be the safest place to stay while waiting for a breeding position to open up elsewhere, or to gain access to quick affairs with members of neighboring groups. Seychelles warblers that delay dispersal on good territories have the highest lifetime fitness. High rates of adult mortality result in a high breeder turnover, which can mean a quick succession for helper superb fairywrens and Seychelles warblers, particularly when the death of both parents removes the need to avoid incest.

LEFT

Sociable weavers breed in "apartments" within a much larger condominium nest. All share the insulation benefits of a thick external roof, but individuals contribute to building this more when they are more related to other inhabitants, and target their thatching to areas that contain their relatives.

Apprenticeships

Seychelles warbler helpers that were relocated to new islands as part of a larger conservation effort took four months to produce a fledgling, which was just as fast as breeding females. In contrast, relocated females that had never bred or helped as babysitters took over a year to fledge any chicks. These inexperienced females built less sturdy nests, which meant they had to spend more time incubating, and had a lower hatching success. Evidence for the benefits of experience at the expense of breeding is limited, as biologists cannot rule out that some of those "helpers" had never snuck some of their own eggs into the nest they were attending.

Babysitting benefits

In contrast to subordinate helpers, dominant members of cooperatively breeding groups almost always benefit from help with babysitting. Among brown-headed nuthatches, stay-at-home sons increase their parents' breeding success by 50 percent. White-browed sparrow-weavers and chestnut-crowned babblers with more helpers also fledge more chicks. In addition to increasing the number of fledglings, carrion crows with helpers fledge heavier chicks.

Load lightening

Helpers can be worth the extra housing and feeding expenses because they lighten the load for breeders. Red-cockaded woodpecker helpers do not increase the amount chicks are fed, but they do decrease individual workloads, and lengthen breeder lifespans. Superb starling and western bluebird helpers have a similarly beneficial effect on the lifespan of breeder females.

LEFT

Leaves are poor fuel for growing chicks and high-energy pursuits like flying because they take a lot of time to digest and provide a paltry amount of energy for their weight. By providing assistance, helpers at hoatzin nests enhance the growth and survival of young and lighten the load for breeding females.

SEX BIASES

Most offspring who stay to help at the nest are sons. In a few species, such as Seychelles warblers, it is mostly daughters, and in fewer cases, such as pied babblers, both sexes stay home.

Among birds, the male bias in dispersal and helping are related. Female birds typically disperse further than males. Females also tend to perform more of the risky jobs, such as incubation. As a result, more females die than males, leading to a shortage of breeding females. When biologists removed breeding male superb fairywrens, helpers filled the vacant positions within hours. In contrast, removing a dominant pair led to an unclaimed vacancy until the female was returned a few days later.

Male helpers stand to gain more than females for two reasons. In the case of western bluebirds, western scrub jays, and any species where males are more likely to live near relatives than females, males are more likely to be related to the chicks they are helping to raise. Not surprisingly, male scrub jay helpers are much more attentive to nestlings than females. In the case of pied kingfishers and other species where helpers may not be relatives, unrelated males still stand a better chance of breeding by joining a group than by trying to breed independently because breeding females are in such short supply.

Manipulating sex ratios

Seychelles warblers benefit from more female helpers in times of plenty, and from more sons, which disperse, when food is scarce. A cross-fostering experiment showed that on good territories, additional helpers improved fledging success, whereas on poor territories, helpers were more of a hindrance. Breeding pairs that were moved to a new island with better territories went from producing 90 percent sons to 85 percent daughters. This dramatic skew in sex ratios is not the result of egg or chick mortality, and seems to be determined by mothers as they make and lay eggs.

LEFT
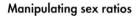

Seychelles warblers are unusual among cooperatively breeding birds because most of their helpers are female.

INCEST AND INFIDELITY

Considering that there are many chances for grown offspring to mate with their parents, incest is almost unheard of among cooperatively breeding birds.

Pied babblers avoid incest because they learn to recognize individual call signatures, but this can result in inadvertent inbreeding between distant relatives who have never met. Superb fairywrens are particularly adept at avoiding incest. Even though social pairs are often first-order relatives such as mothers and sons, they almost never mate with each other, and if they do, their chicks are stunted.

As superb fairywrens have the highest recorded infidelity rates among birds, the dominant breeding male has the shortest straw of all group members. Even when he isn't related to his social mate, more than three-quarters of her chicks are not fathered by him, but by males from other groups.

Helpers also liberate the dominant female from having to remain faithful as a way of ensuring her social mate performs his share of childcare. To add insult to injury, he also loses extra-group copulations to these subordinate males because neighboring females often end up mating with a hanger-on instead of the top male who has been assiduously courting her for months.

LEFT

Breeding male superb fairywrens display to neighboring females (but not to their social mates) with flower petals. The best option for a dominant male superb fairywren is to father as many extra-group offspring as possible, since he won't have to devote any effort to caring for them.

COALITIONS AMONG NONRELATIVES

Rather than establishing family clans, some species form groups of unrelated adults sharing a communal nest. To persist, these groups must benefit their members more as a collective than as solo breeders. As individuals stand no chance of getting their genes into the next generation by helping nonrelatives at the expense of their own reproduction, unrelated groups are far more egalitarian than family clans, and usually, everyone gets to breed.

However, birds that breed with nonrelatives experience more conflict within groups. Nest robbery, egg desertion, and infanticide are all much more common among species like acorn woodpeckers, Mexican jays, and anis, that nest with nonrelatives. Cooperation among unrelated dunnocks is largely the byproduct of conflict, when two males sharing parental duties at a single nest reach an impasse, where neither has succeeded in monopolizing a female, but both have genetic stakes invested in a shared nest.

In contrast, male Galapagos hawks cooperate in a peaceful coalition that brings net benefits to all its members. This could be because the males have no idea who has actually fathered the chicks. Nevertheless, coalition members share territory defense, and larger groups are able to occupy larger territories, which in turn increases their chances of living to a ripe old age. Even if each coalition member has few offspring per year, coalition membership brings a net benefit in terms of lifetime reproduction because of increased longevity. This pattern is true for the handful of other species that nest with nonrelatives, including riflemen and pukekos of New Zealand, green woodhoopoes, and Taiwan yuhinas.

TOP LEFT

Mexican jays breed in groups, but experience high levels of within-group conflict.

LEFT

Riflemen from New Zealand are an example of relatively peaceful cooperative breeding between nonkin.

ACORN WOODPECKERS

Acorn woodpeckers in northern California live and breed in large groups that communally amass and defend granaries of acorns individually stored in holes the woodpeckers have drilled into tree trunks. In contrast, populations in New Mexico breed in smaller groups, while migratory populations in southeastern Arizona neither breed jointly nor store acorns. Acorn woodpeckers are unusual among cooperative breeders, as larger groups do better in times of plenty. This is because larger groups are better able to defend their granaries.

From mate sharing to nest sharing

In contrast to cooperative polyandry, in which males share a single female and cooperate to care for young at a single nest, joint nesting, in which females share a nest, is the rarest form of cooperative breeding, recorded for fewer than 20 species. Even rarer is joint nesting with nonrelatives.

About 2–3 percent of female oystercatchers are cooperatively polygynous, sharing a nest and all offspring care. Typically one of these females is a widow making the best of a bad job. There is no evidence that males benefit from polygyny, as two females at a nest hatch fewer chicks than a single female. Like many other cooperatively breeding females, these oystercatchers copulate and preen each other regularly, and watching biologists are unable to distinguish between these homosexual encounters and heterosexual ones. One explanation for homosexual copulations is that it promotes peace and pair-bonding. However, some joint nesters, like acorn woodpeckers, engage in plenty of within-group conflict while also copulating and cooperating with members of the same sex.

Other joint nesters, such as white-winged choughs and pukekos, live in peaceful polygynandrous communes without any socially monogamous pair-bonds tying couples together. This sexual freedom could be the result of males having no time to guard any particular female because they have to devote much of their time to defending the group territory.

Keeping the peace

Pukekos are bright purple New Zealand relatives of moorhens that typically breed in groups with two or more females that show no signs of squabbling, and frequent female–female copulations. Like most female joint nesters, female pukekos face the problem of incubation constraints. The larger the group, the more eggs there are to incubate, and the fewer each female hatches successfully. This is partly because incubating a large clutch is inevitably going to leave some eggs out in the cold, and because more eggs get broken as parents change incubation shifts for a larger clutch. As a result it would benefit females to remove each other's eggs, but, puzzlingly, they don't.

What keeps the peace between female pukekos? Biologists wondered if females refrain from destroying each other's eggs for fear of retaliation. Although experimentally removing eggs failed to spur females to punish their cobreeders, it did prompt the males to abandon the nest in favor of starting again. Male pukekos invest more in parental care than females, including the more dangerous night incubation shift and most chick-rearing duties, so they provide the check on any egg sabotage between females.

TOP LEFT

It takes two adults to raise oystercatcher chicks, so female oystercatchers sometimes team up to rear chicks together.

LEFT

Pukeko females lay their eggs in a joint nest, and show very few signs of competition, while male pukekos do most of the chick rearing.

COOPERATION AND COMPETITION

Even though reciprocity should constrain groups of unrelated individuals to distribute reproductive benefits equally, individual members can vie for a larger share of the collective profits. Surprisingly, equality among joint nesters can be maintained by either reduced or increased competition within the group.

Harsh weather promotes cooperation among Taiwan yuhinas, making it worthwhile to share a nest with nonkin. Larger groups are better buffered against bad times because they can renest faster and because group breeding lengthens individual lifetimes. In addition, larger groups are best able to lower predation by making fewer commotions during feeding visits to the nest. Group members achieve this by synchronizing their visits so that several members visit the nest at once, rather than having to make lots of individual trips to deliver the same amount of food.

ABOVE

We know very little about the breeding habits of birds from the Far East, but Taiwan yuhinas are a fascinating exception. A Japanese officer first described joint nesting in Taiwan yuhinas in 1938, when he noticed 4–8 adults all tending the same nest. He also inferred from the presence of distinct egg patterns, that 2–3 females contributed eggs to the communal clutch.

LEFT

Among joint nesting woodpeckers and four species of New World cuckoo, including these guira cuckoos, intensely wasteful conflict between cobreeding females has the rather harmonious effect of enforcing reproductive equality.

Unlike the openly promiscuous pukekos, Taiwan yuhinas form socially monogamous pairs. Both sexes establish a pecking order, and females compete to have the most eggs in the communal nest. Females tussle by landing on the back of whoever is in the act of egg laying, pecking and attempting to shove her off the nest. When there is plenty of food, dominant females tend to win the most tussles, lay their eggs earlier, and initiate nocturnal incubation. The last eggs laid have the smallest chance of surviving because they hatch last, if at all. However, when times are hard, tussles are less intense, and everyone lays fewer eggs, resulting in greater reproductive equality between dominant and subordinate females.

Sabotage among joint nesters

Guira cuckoos and greater, groove-billed, and smooth-billed anis are all joint nesting species of cuckoo, with females that do their best to sabotage each other's breeding attempts, to make more room for their own eggs. Having too many eggs in a clutch reduces incubation efficiency, so a smaller proportion hatch, which is bad for everyone.

Females that are first to lay lose the most eggs to other females, as they stop throwing out eggs once they've started laying their own, so as to avoid killing their own young by mistake. Females that start laying last can spend the most time sabotaging others without risking their own eggs. Female smooth-billed anis nesting alone or with one other female lay an average of five to six eggs a season, but early layers in large groups of four to five females have to lay as many as thirteen to compensate for the greater amount of egg tossing.

From cooperation to parasitism

There is a very fine dividing line between joint nesting, where multiple pairs care for eggs in a communal nest, and brood parasitism, in which some breeders simply offload their eggs on more caring individuals and escape all the work of parental care. Strikingly, brood parasitism has evolved three times independently in cuckoos, but four New World cuckoo species happen to be among the minority of cooperative joint nesters.

ABOVE LEFT

Greater anis from Panama are at the most egalitarian end of the continuum. Breeding members are the least likely to tolerate reproductive inequalities because, of all three ani species, they are the least related to other group members.

ABOVE

Guira cuckoo groups contain a mixture of relatives and nonrelatives, with males being more closely related than joint-nesting females. Although guira cuckoos frequently eject eggs to trim down the communal clutch, their exquisite egg patterns are not signatures, and cannot help females to avoid killing their own offspring.

SEVEN

BROOD PARASITISM

Some birds leave all the work of nesting, incubating eggs, and raising chicks to others. Coots and ducks often dump the odd egg into another pair's nest, whereas other birds, like many cuckoos, have evolved to rely exclusively on parasitizing the nests of other species.

PARENTING BY PROXY

Imagine raising a child that begins to grow several times larger than yourself, while somewhere along the way, all your other children have mysteriously vanished, leaving this monstrous, irresistible, and insatiable individual monopolizing all your attention. About 1 percent of living bird species have evolved to be purely parasitic, relying entirely on the efforts of other (typically smaller) host species to make a nest, incubate their eggs, and raise their chicks, and cuckoos and cowbirds are famous instances of this strategy.

RIGHT

Brown-headed cowbirds of North America are an independently evolved instance of brood parasitism.

Partial parenting

Although the whole point of being a brood parasite is to avoid spending time and energy raising chicks, a few species have been observed performing some elements of parental care.

Striped cuckoos of South America have been seen foraging with juveniles, engaging in complex duets with them, and guarding them from predators. Two African cuckoo species have been observed feeding juveniles of the same species. Taken together with occasional parasitism by generally nonparasitic species like the black-billed cuckoos of North America, these occasional parental behaviors by some cuckoos suggests that brood parasitism evolved by degrees. Not *all* brood parasites are equally parasitic.

Playing dirty

Some cuckoos and cowbirds monitor host nests that they have parasitized, and punish any rejector hosts by destroying all the host's own young and forcing the host family to start all over again. This sabotaging behavior is sometimes called "farming," because brood parasites can get a new nest to lay in sooner if they destroy a nest that either no longer contains their eggs, or is already too advanced in the breeding cycle for a parasitic chick to thrive.

LEFT

One of the most iconic images of a reproductive cheat is a common cuckoo fledgling being fed by a tiny and unrelated host parent, like this reed warbler, who almost vanishes into the gaping mouth of the parasitic chick in its nest.

RIGHT

Striped cuckoos are New World cuckoos that independently evolved to become obligate brood parasites of other species.

PARASITISM WITHIN SPECIES

Although parasitizing other species is relatively rare, many bird species occasionally dump their eggs in the nests of other members of their own species. The two forms of parasitism seem largely unrelated. Birds that sometimes parasitize their own kind have seldom evolved to rely on a different species to nest, incubate eggs, and raise chicks for them.

LEFT AND ABOVE

Many birds, including the European starling (left), occasionally lay their eggs in the nest of another member of the same species.

There's a fine line between accepting help with childcare and manipulating others into doing all the care for you. At what point does using nannies, especially unrelated ones, become exploitation rather than cooperation, and can one behavior lead to the evolution of the other?

Some group breeders, such as white-fronted bee-eaters, have helpers at the nest, but also lay their eggs in the nests of other females. European starling females sometimes "dump" an egg in the nest of another female, but they also remove one of the host's eggs. Parasitizing members of the same species has been reported in over 230 birds and the number is likely to increase as more genetic parentage studies are done.

Why would a female decide to lay her eggs in someone else's nest? Parasitizing one's own kind is most common among species that nest in large groups and when the cost of raising a few foundlings is relatively small. When a female has lots of offspring, all of whom are relatively independent and require minimal care, it can pay to sneak a few eggs into a relative's nest rather than raising all her young herself or accepting helpers at the nest. This is because the minor cost to foster mothers of raising a few extra young is conveniently offset or even exceeded by the evolutionary benefit of raising relatives. A bird

can get more genes into the next generation not just through its own offspring, but also through nieces, nephews, brothers, and sisters. In most ducks, females that nest near each other are very likely to be related because daughters tend to stay near their place of birth. Their independent ducklings require little parental care, and within-species parasitism is found in many duck species.

In contrast, songbirds and other species with naked nestlings require a great deal of parental care, so raising someone else's young is extra costly. Parasitized cliff swallow parents have to waste a lot of time and effort raising helpless and demanding chicks in which they have no genetic investment, shortening their lives and reducing the number of biological offspring they can afford to rear.

ABOVE

It is relatively cheap for waterfowl like these common merganser to raise a few extra highly independent and precocial offspring even if they have no genetic stake in them.

These house sparrow eggs have each been laid by a different female, and sport distinct signatures.

When the costs of being parasitized are high, one would expect to see defenses evolve to minimize the chances of rearing another pair's offspring. Weaverbirds and house sparrows can recognize their own eggs and reject those of another female, and zebra finches identify their own eggs by smell. In these cases, laying an egg in someone else's nest is often a female's way of making the best of a bad job, such as when her nest is destroyed and she has to lay an egg somewhere. Whether or not to dupe members of your own species into raising some of your offspring is a conditional, optional strategy, in contrast to the obligate brood parasites, like cuckoos, which have lost the ability to make nests or do any parenting, and rely purely on the efforts of other species to incubate their eggs and raise their chicks.

Backyard Bird

AMERICAN COOTS

American coots often sneak their eggs into another coot's nest, and this is very costly for the hosts because they typically lose one of their own offspring for every successful parasitic chick. As a result, coots have evolved many ways to minimize being duped into caring for someone else's young. Coots learn to reject eggs that don't resemble their own, and actually recognize and kill parasite chicks by imprinting on the very first chicks that hatch. This strategy works because parasitic attempts always take place after the host female has started laying, and coots begin incubation with their first egg, so parasite chicks always hatch later.

OBLIGATE BROOD PARASITISM

Some brood parasites, like the common cuckoo, are specialists, in the sense that a single female is exquisitely adapted to lay her eggs in the nest of a single host species. Others, such as the brown-headed cowbird of North America, are generalist parasites, with individual females laying their eggs in the nests of several species, from blackbirds to bluebirds, even in a single breeding season.

RIGHT

Honeyguide chicks are born with sharp bill hooks to destroy their host's own chicks.

Variation in virulence

Just like any parasite, avian brood parasites vary in how virulent they are. Honeyguides and striped cuckoos hatch with special bill hooks that they use to stab and worry all the other chicks and eggs around them, while other newly hatched cuckoo species blindly eject all the host's eggs and chicks from the nest. Other parasites like the great spotted cuckoo, African *Vidua* finches, and cowbirds have no evicting or killing instincts upon hatching, but some of them will outcompete the host chicks they grow up with and indirectly starve the competition to death. Some brood parasites give their chicks a head start by incubating their eggs internally for an extra day, which can enable the larger parasite chick to hatch before its nest mates and begin killing them either directly or indirectly.

LEFT

Little bee-eater chicks do not always reach the stage shown here—they are parasitized by greater honeyguides, whose newly-hatched chicks will peck host hatchlings to death.

Why isn't obligate brood parasitism more common?

One might expect to see many birds evolve to exploit the parental efforts of others, freeing them to invest in laying more eggs than species that have to devote energy to childcare. As Darwin pointed out, the parasitic lifestyle also allows those that breed in temperate regions, like the common cuckoo, to migrate back to the balmy tropics earlier than other birds that stay until autumn raising chicks. But brood parasitism is comparatively rare among the approximately 10,000 species of birds. Obligate brood parasitism occurs in only 1 percent of all birds, and has evolved independently seven times: once each in the cowbirds and estrildid finches, once in the honeyguides, three times within the cuckoos, and once in ducks.

Being parasitized is costly for the hapless host parents, so they fight back, which could account for why reproductive cheating is less common than you might expect. Also, not all bird species are equally ripe for exploitation. Some may live on a diet unsuitable for a young parasite, or nest in inaccessible places, or simply be too large for a parasite chick to compete against.

BELOW

The black-headed duck of South America parasitizes a range of other waterfowl and, occasionally, gulls or birds of prey.

Where brood parasites live and why

The highest concentrations of brood parasitic species are in sub-Saharan Africa and Australasia. Curiously, this geographic distribution corresponds strikingly well with that for cooperative breeders. Is this just a chance correlation, or have brood parasites evolved to target cooperative breeders? Horsfield's bronze-cuckoos of Australia benefit from targeting a cooperative breeder, the superb fairywren, because cuckoo chicks raised by larger fairywren groups grow faster and are more likely to survive to fledge as larger groups bring in more food and are more vigilant against nest predators.

There is a twist to this, however, because larger groups of fairywrens are also better defended against cuckoo parasitism. Superb fairywrens have a specific alarm call for "cuckoo," which they use to mobilize group defenses, and larger groups spend more time mobbing cuckoos, are less often parasitized, and successfully raise more of their own offspring than smaller groups. So while cuckoos may benefit from parasitizing cooperative breeders if they can get through a group's defenses, they are simultaneously selecting for hosts to breed in larger groups so as to better guard against parasitism. Being the target of brood parasites like cuckoos could be both a cause and a consequence of cooperative breeding in hosts.

ABOVE

Diederik cuckoos, named for the sound of their plaintive call, breed across most of Sub-Saharan Africa and the Arabian Peninsula, and primarily parasitize a range of species in the weaverbird family.

LEFT

Global patterns of species richness during the breeding season show a striking similarity between bird species that breed cooperatively (see page 140) and obligate avian brood parasites (left). This is after controlling for the fact that there are more species near the tropics, which means that a higher proportion of all bird species are cooperative breeders or brood parasites, especially in sub-Saharan Africa and Australasia. The darker the shading on the map, the more common brood parasites are.

171

AN EVOLUTIONARY ARMS RACE

Brood parasites that rely on others for all aspects of parental care are locked in an arms race with their hosts. Being parasitized is costly, so host birds are constantly responding to new parasitic tricks by coevolving defenses. One form of parasite trickery is to specialize in mimicking the appearance of the eggs of a particular host species. In turn, hosts may become increasingly adept at distinguishing between parasite eggs and their own.

BELOW

Common cuckoos have more than 100 host species whose nests they lay their eggs in, although individual female cuckoos specialize.

Evolution of egg mimicry in the common cuckoo

Common cuckoos are specialist brood parasites, with individual females targeting particular host species that will raise their young. Female cuckoos specializing on naïve hosts need not lay mimetic eggs to get their eggs accepted. However, females from lineages that have been coevolving with hosts for longer must lay eggs that are better forgeries. Hosts counteradapt with defenses that increase their ability to detect cuckoo eggs, such as evolving egg pattern signatures that are harder to copy.

The various stages of a coevolutionary arms race between specialist brood parasites and their hosts are beautifully illustrated by the evolution of egg mimicry by common cuckoo lineages. In each case the cuckoo egg is shown on the left and the host egg on the right. What may look like the best match to the human eye differs for birds, who also see in ultraviolet.

1. Early in an arms race, naïve hosts like dunnocks have seldom paid the price of incubating parasite eggs in their evolutionary past, so they are most undiscriminating and will incubate mismatched cuckoo eggs. Although cuckoos have the ability to evolve matching blue eggs to avoid rejection by more discriminating hosts in Europe and China, these dunnock-specialist cuckoos simply haven't needed to.

2. Meadow pipits and the cuckoo lineage specializing on them have been coevolving for longer, so meadow pipits do reject some mismatched eggs, and only the cuckoo eggs that are reasonably good forgeries survive.

5. Reed buntings lay hard-to-forge egg signatures and probably emerged as winners of an old arms race against cuckoos. Nowadays, reed buntings are rarely parasitized, and there is no cuckoo race that specializes on them, but they still reject all mismatched eggs experimentally placed in their nests, suggesting they have retained the ability to recognize and reject cuckoo eggs from their evolutionary past and that the cuckoo lineage that once specialized on reed buntings has gone extinct or switched to a different host species.

4. Red-backed shrikes and their cuckoos are probably close to the end of their arms race. The shrikes reject almost all mismatched eggs, and have evolved egg signatures with unevenly distributed speckles that are harder for cuckoos to forge than the evenly mottled meadow pipit eggs.

3. Reed warblers have been locked in an even older arms race with their specialist cuckoo lineage, and have evolved excellent egg discrimination abilities, in turn selecting for better egg forgeries by the cuckoos that parasitize them.

173

Why don't hosts reject monstrous and clearly mismatched parasite chicks if they are able to achieve sophisticated levels of discrimination between a foreign egg and their own? There is no clear answer to this question, but most explanations rely on the fact that hosts can mount more than one line of defense, each of which can be breached by parasite counteradaptations, which would then force hosts to evolve a new line of defense. Sometimes, the costs of maintaining a complete armory outweigh the benefits of being fortified at one or two points where parasites can strike.

Defense at the front line

The first line of defense, and the best for a host, is at the nest. If a host can stop a parasite from laying an egg in its nest, it can avoid all the costs associated with trying to identify and reject a foreign egg. This is especially cost-effective if the parasite females are "egg-removers," taking an egg each time they lay in a host nest. However, the front line can be tricky to defend because female parasites will spend a lot of time discreetly scouting out potential host nests so they can insert an egg at just the right time, when a host is in the middle of completing her own clutch. Once they've picked a target, all obligate brood parasites, with the exception of the black-headed duck, can lay their eggs in a matter of seconds, which enables a quick getaway.

LEFT

Baya weavers in Asia construct nests with long entrance tubes, which could minimize parasitism attempts, while also restricting access by predators.

The parasitic cuckoo finch female (left) of sub-Saharan Africa looks just like a harmless female red bishop (right), which confuses prinia hosts and reduces mobbing attacks.

TOP

Some hosts, such as meadow pipits, recognize and mob a female cuckoo to stop her from laying in their nest.

Hosts can make it hard or impossible for a parasite to find or enter their nests. For instance, some weaverbirds construct nests with entrance tubes about a foot long, in which cuckoos can get stuck and are vulnerable to the mobbing attacks of their hosts. Many birds also guard their nests, making it difficult for a parasite to lay an egg or even to monitor the host nest so she can time her egg laying correctly. For instance, cape bulbuls from Africa are so good at attacking Jacobin cuckoos that cuckoo eggs are often laid too late and fail to hatch. However, the bulbuls don't seem to reject cuckoo eggs or chicks that do make it past that very strong first line of defense.

In response to these frontline defenses, brood parasites have evolved some rather sophisticated tricks of their own. Some cuckoo species have evolved to look like sparrowhawks, major predators of small birds, making it hard for their hosts to know if they should fly for cover or start attacking a snooping cuckoo. Male great spotted cuckoos and Asian koels will distract host parents so that while they are mobbing him, his mate can sneak in and deposit her egg in their nest.

Defense at the egg stage

Hosts can escalate defenses at the egg stage by evolving egg signatures. At the simplest level, this can be achieved by increasing the differences in egg appearance between females while reducing variation within a clutch. Maintaining these signatures must be costly because host populations freed from parasitism lose the variation,

BELOW AND BELOW RIGHT

Common cuckoos (left) mimic sparrowhawks (right) to reduce mobbing by host bird species.

rather like birds on islands like the Galapagos losing their instinctive fear of predators. African village weavers were introduced to the Caribbean in the late 1700s, and in the absence of their native cuckoo, the eggs of individual females are less distinct from each other and less uniform within a clutch, rather like one's signature getting less unique and less repeatable when the risk of forgery is reduced.

A particularly elegant example of escalation at the egg stage takes place in Zambia between several specialist races of the cuckoo finch and three of its hosts. Like the common cuckoo, female cuckoo finches each specialize on a particular egg color and pattern that mimics a particular host's egg. To make it harder for the cuckoo finch to match their eggs, tawny-flanked prinias have evolved a fantastic array of egg signatures comprising multiple components that can be mixed and matched, so while one female consistently lays pale red eggs with a few splotches all concentrated at one end of the egg, another may lay an egg of similar color but evenly specked or with smaller or more speckles. Cuckoo finches specializing on these prinias have evolved to match almost every aspect of these egg signatures, but because each female can only lay a certain signature, and lays at random in the nests of different prinia females, their eggs are often mismatched and rejected. Some prinias have evolved an olive-green egg that the cuckoo finches have yet to match, and this color has become more common in a highly parasitized prinia population in the space of just 40 years.

None of the other host species parasitized by cuckoo finches have evolved anything close to this level of variation in egg appearance, but they are more discriminating than the prinias. Indeed, there appears to be a trade-off between variation and discrimination, in which the host species with the least egg variation shows the strongest egg rejection behavior, whereas the prinias are the least likely to reject foreign eggs.

TOP RIGHT

Parasitic cuckoo finch eggs laid by different females (inside circle) match most of the egg signatures laid by their host, the tawny-flanked prinias (outer circle), with the exception of the olive-green eggs at 9 o'clock.

RIGHT

Individual tawny-flanked prinias are genetically programmed to lay eggs of a particular pattern and color, but the tremendous variation across individual egg signatures within this species makes it especially hard for parasitic cuckoo finches to lay good egg forgeries.

Backyard Bird

PIN-TAILED WHYDAHS

Parasitic pin-tailed whydah chick gapes (left) look almost identical to the gapes of their hosts, the common waxbill (right). This was thought to be another case of chick mimicry by parasites, but there is little evidence that hosts reject *Vidua* parasite chicks with mismatched mouthparts. Instead, the parasite has probably evolved elaborate mouth markings to stimulate host parents to feed them more, which in turn has forced the host's own chicks to keep up by evolving to mimic that parasite's attractive mouth patterns.

RIGHT

Examples of excellent chick mimicry by cuckoos include the Australasian bronze-cuckoo (top left), whose chicks mimic both the skin color and some of the fuzz on their host's chicks (large-billed gerygone, top right); the shining bronze-cuckoo (middle left) and its host the yellow-rumped thornbill (middle right); and the Horsfield's bronze-cuckoo (bottom left) with its host the superb fairywren (bottom right).

Defense at the chick stage

For a long time, biologists puzzled over the lack of chick rejection as a defense, and the most plausible explanation was that since hosts have to learn how to reject parasitic eggs by imprinting on their own, they would have to apply the same strategy to detect foreign chicks. This would be more difficult with chicks, which, unlike eggs, change appearance constantly as they grow, and could backfire badly if first-time parents were parasitized, had all their own young destroyed by the cuckoo, and learned to accept cuckoo chicks and reject all their biological offspring for the rest of their lives.

However, it turns out that chick rejection does exist, even among hosts of "evictor" parasites like the bronze-cuckoos of Australia, and that this line of host defense has selected for chick mimicry by some parasites. These hosts don't reject at the egg stage, but will either abandon their nest when they see a cuckoo chick hatch, or pick up a newly hatched cuckoo chick and fling it out of the nest before it has a chance to start ejecting their eggs. In South America, bay-winged cowbird hosts will feed mimetic screaming cowbird fledglings but stop feeding nonmimetic shiny cowbird fledglings. In addition to looking like host chicks, some brood parasites have evolved to sound like host chicks; others even get more food by sounding like a complete brood of host chicks.

Egg passwords

How does a group of cooperatively breeding superb fairywrens know to reject visually mimetic cuckoo chicks? The solution is for females to teach their chicks a secret "password" through the eggshell by including unique elements in their incubation call. Females also teach this vocal signature to their mate and to helpers at the nest.

While still in the egg, fairywren embryos can distinguish individuals of their own species. The more incubation calls they hear through the egg, the better fairywren hatchlings are at matching the signature elements in their begging and the more food they get from all the adults in the group, all of whom know the password.

A cuckoo has less time to learn the password because it hatches days before the fairywrens so it can toss all the fairywren eggs overboard and monopolize the nest. Instead, the cuckoo relies on trial and error to tailor its begging calls to a particular host, and can only do so after hatching, so those that fail to match the host female fast enough are abandoned.

ABOVE

The Horsfield's bronze cuckoo breeds in Australia and winters in Indonesia, Malaysia, and Singapore. These cuckoos parasitize many host species, most of which make covered nests.

LEFT

Female superb fairywrens repeat a password to their chicks through the eggshell. Cuckoo chicks, who fail to learn this password, are rejected.

ESCALATE OR TOLERATE?

When one looks at the sophisticated trickery and defense that has evolved between brood parasites and their hosts, it is easy to assume that undiscriminating hosts are lagging in the evolutionary arms race. An alternative explanation is that resistance can be too costly. Instead of escalating, hosts are sometimes better off passively accepting cuckoos, or can actively evolve tolerance to offset the cost of being parasitized.

When resistance is futile or too costly, passive acceptance can be the best option. South American shiny cowbirds typically puncture one or more eggs in every nest they parasitize, and their chalk-browed mockingbird hosts will vigorously defend their nest, sometimes pecking at a cowbird's eyes even as she is inside it. In spite of their strong frontline defenses, the mockingbirds are often parasitized by multiple female cowbirds and accept all the cowbird eggs. It turns out this acceptance leads to more surviving host eggs because having multiple cowbird eggs actually dilutes the clutch, lessening the chances that the mockingbird's eggs are the ones destroyed by a cowbird.

ABOVE

Chalk-browed mockingbirds are common hosts of the shiny cowbird.

RIGHT

Like the brown-headed cowbird of North America, shiny cowbirds do not destroy all the host eggs in a nest.

Acceptance can lead to the evolution of tolerance to compensate for being parasitized. Host parents can hedge their bets by having more, smaller clutches in a season in the hope that not all of them will be lost to parasitism. Hosts with a longer history of being parasitized by the North American brown-headed cowbird lay more small clutches than those more recently exposed to cowbird parasitism. Conversely, Montezuma oropendolas typically lay an insurance egg which gets tossed out in the absence of parasitism.

Great spotted cuckoos and brown-headed cowbirds also keep an eye on nests that they have parasitized and prey on the nests of hosts that reject their eggs, so passive acceptance or tolerance could be better options than resistance in the face of these tactics.

LEFT

Montezuma oropendolas have adapted to tolerate brood parasites by laying an extra egg as insurance against destruction by a parasite.

Backyard Bird

GREAT SPOTTED CUCKOOS

In an interesting twist, some of the less virulent parasites and their hosts have gone from tolerance to mutualism. Having a great spotted cuckoo in the nest actually benefits crows in two ways. The cuckoo chicks produce a special repellent that keeps nest predators away, and crow nestlings actually beg less and free-ride on the vigorous begging efforts of the cuckoo chick while conserving their own energy for growth.

FLEXIBLE, LEARNED RESPONSES

Smart hosts learn a range of responses to the threat of parasitism. Eurasian magpies, for example, who are subject to punishment if they reject a parasite's eggs, learn to accept cuckoo eggs in response to these mafia tactics.

Magpies are known to be clever, but even small songbird hosts will learn to defend themselves flexibly against parasites when the threat of parasitism is high. Reed warblers in Europe learn to recognize and mob cuckoos by watching their neighbors, and increase their overall vigilance when they see and hear their neighbors attacking cuckoos. Similarly, more experience leads to better egg discrimination, and hosts with the ability to reject parasite eggs primarily do so not by spotting the odd egg in a nest, but by learning what their own eggs look like and rejecting any that look sufficiently different.

TOP LEFT

In experiments, Eurasian magpies choose to nest farther away from places they perceive to be at high risk of parasitism where biologists consistently play cuckoo calls and put out stuffed cuckoos to simulate areas of high cuckoo density.

FAR LEFT AND LEFT

Female common cuckoos come in two color morphs. The gray morph (left) is a better mimic of a common bird predator, the sparrowhawk, and reed warbler hosts only learn to mob this morph instead of fleeing from danger after watching their neighbors mobbing it. The rufous morph (far left) can reduce mobbing by being less common than the gray morph and so less easily learned and recognized as a cuckoo.

EVOLUTIONARY EGG RACES

How do specialist parasites like common cuckoos maintain the appropriate adaptations to manipulate their particular host species? Some of these adaptations, like where to lay one's egg, are probably learned, whereas responding to the correct alarm call is a combination of an innate template and experience. The color and pattern of eggs an individual female lays are genetically determined and barely altered by the environment.

One evolutionary route to take is to speciate, so that everyone raised by a particular host species interbreeds, but remains reproductively isolated from parasites raised by other hosts, so that the genetic changes that allow for some of these complex and specialized adaptations like the color and degree of egg speckling remain distinct rather than being mixed and matched every generation.

Rapid speciation within the parasitic African *Vidua* finches has led to each species specializing on a single host. Unlike other obligate brood parasites, these finches parasitize closely related species and males actually incorporate parts of their host male's song into their own repertoire, while females learn who to mate with by imprinting on their host male's song. This effectively produces a new *Vidua* species every time a few females mistakenly lay their eggs in the nests of a new host species, because all their offspring will only mate with other individuals that sound like the new host, keeping them genetically isolated from the original lineage.

In contrast, there is mixed genetic evidence that highly specialized races of common cuckoo are evolving into distinct species. Many of these races have more than one origin, and there is genetic mixing between females and males raised by different hosts. Yet we know that cuckoo females from different lineages have some exquisite adaptations to mimic host eggs that probably require a suite of genetic differences.

LEFT

Pin-tailed whydahs parasitize closely related estrildid finches, and incorporate elements of their host's song into their own.

CUCKOOS AND COWBIRDS IN CONSERVATION

Just like any species, including humans, generalists like the brown-headed cowbird of North America are often better able to respond and adapt to, and even take advantage of, rapidly changing environments, whereas specialists like the common cuckoo are facing greater difficulties in the face of rapid change. Human-induced alterations to the environment, from misguided management efforts to climate change, have threatened both brood parasites and many of their hosts.

Brown-headed cowbirds are named for their habit of following bison (and subsequently cattle) herds to feed on the insects flushed by the grazing mammals. Originally restricted to the Great Plains, this parasite began spreading throughout the United States in the 1800s, as humans opened up new habitat for it by clearing swathes of forest. As generalists, brown-headed cowbirds have parasitized more than 200 species, many of which are especially vulnerable because they have not had any time to evolve defenses.

This has led to some species, such as the Kirtland's warbler, coming close to extinction. These birds declined by 60 percent in a decade due to a combination of cowbird parasitism and habitat loss. In addition to being naïve hosts for increasing numbers of cowbirds spreading into their native range, this warbler is extremely exacting about nesting in small jack pines, and decades of fire suppression by humans has reduced the number of suitable trees available.

LEFT

The rapid spread of brown-headed cowbirds throughout North America has followed the clearing of forests by humans.

There is a happy ending to this story for the warblers, if not for the cowbirds, as prescribed burns and systematic cowbird extermination have rescued the Kirtland's warbler population from about 400 breeding pairs in 1971 to more than 2,000 pairs today.

The common cuckoo itself is getting rather rare, at least in some parts of the world. Once the acknowledged harbinger of spring, British cuckoos have declined by over half in just the last 25 years. Climate change has affected the migratory patterns of many birds, and some satellite-tracking studies of common cuckoos show that they often leave their breeding grounds to return to Africa as early as the start of June, when their hosts are still busy breeding. The extremely short time spent in their breeding range could make cuckoos especially vulnerable to climate change altering the best times for them to migrate. An international project is now tracking cuckoos from China, Belarus, Germany, and the UK to their wintering grounds in Africa, where they spend most of their lives.

In the meantime, the cuckoo decline is good news for their hosts in the UK. In 1985, almost a quarter of reed warbler nests had cuckoos in them, but by 2012 this had dropped to a mere 1 percent. Fascinatingly, reed warblers have responded by rapidly losing their anticuckoo defenses. The proportion of reed warblers mobbing cuckoos (or experimental cuckoo models) has declined from 90 percent to 38 percent, and the proportion rejecting mismatched model eggs has declined from 61 percent to 11 percent.

TOP RIGHT AND RIGHT

Many small hosts, such as Kirtland's warblers (top), accept cowbird eggs because they are unable to remove them without damaging their own. But yellow warblers (right) rebuild over a clutch containing a cowbird egg. As cowbirds often return to the same pair, this can result in a multistory nest of clutches that will never hatch.

FURTHER RESOURCES

Books

Black, J. M. (1996). *Partnerships in Birds; the Study of Monogamy*. Oxford University Press.

Davies, N. B. (2000). *Cuckoos, Cowbirds and Other Cheats*. T. & A. D. Poyser.

Davies, N. B. (1992). *Dunnock Behaviour and Social Evolution*. Oxford University Press.

Davies, N. B., Krebs, J. R. & West, S. A. (2012). *An Introduction to Behavioural Ecology*. 4th edition, Wiley-Blackwell.

Goodfellow, P. & Hansell, M. H. (2011). *Avian Architecture: How Birds Design, Engineer & Build*. Ivy Press.

Koenig, W. D. & Dickinson, J. L. (Ed.) (2016). *Cooperative Breeding in Vertebrates: Studies of Ecology, Evolution, and Behavior*. Cambridge University Press.

Koenig, W. D. & Dickinson, J. L. (Ed.) (2004). *Ecology and Evolution of Cooperative Breeding in Birds*. Cambridge University Press.

Payne, R. B. & Sorenson, M. D. (2005). *The Cuckoos*. Oxford University Press.

Stacey, P. B. & Koenig, W. D. (1990). *Cooperative Breeding in Birds: Long-term Studies of Ecology and Behavior*. Cambridge University Press.

Journal articles

Albrecht, T., Vinkler, M., Schnitzer, J., Poláková, R., Munclinger, P. & Bryja, J. (2009). Extra-pair fertilizations contribute to selection on secondary male ornamentation in a socially monogamous passerine. *Journal of Evolutionary Biology*, 22, 2020–2030.

Alonso-Alvarez, C., Doutrelant, C. & Sorci, G. (2004). Ultraviolet reflectance affects male-male interactions in the blue tit (*Parus caeruleus ultramarinus*). *Behavioral Ecology*, 15, 805–809.

Amundsen, T. (2000). Why are female birds ornamented? *Trends in Ecology & Evolution*, 15, 149–155.

Andersson, Malte. (2005). Evolution of Classical Polyandry: Three Steps to Female Emancipation. *Ethology*, 111(1), 1–23.

Aznar, F. J. & Ibáñez-Agulleiro, M. (2016). The function of stones in nest building: the case of Black Wheatear (*Oenanthe leucura*) revisited. *Avian Biology Research*, 9(1), 3–12.

Berg, K. S., Delgado, S., Cortopassi, K. A., Beissinger, S. R. & Bradbury, J. W. (2007). Vertical transmission of learned signatures in a wild parrot. *Proceedings of the Royal Society B*, 279, 585–591.

Bonato, M., Evans, M. R. & Cherry, M. I. (2009). Investment in eggs is influenced by male coloration in the ostrich, *Struthio camelus*. *Animal Behaviour*, 77, 1027–1032.

Boucaud, I. C. A., Mariette, M. M., Villain, A. S. & Ementine Vignal, C. L. (2015). Vocal negotiation over parental care? Acoustic communication at the nest predicts partners' incubation share. *Biological Journal of the Linnean Society*, 117, 322–336.

Braun, A. & Bugnyar, T. (2012). Social bonds and rank acquisition in raven nonbreeder aggregations. *Animal Behaviour*, 84, 1507–1515.

Brennan, P. L. R. (2012). Mixed paternity despite high male parental care in great tinamous and other Palaeognathes. *Animal Behaviour*, 84, 693–699.

Burg, T. M., Croxall Burg, J. P. & Burg, T. M. (2006). Extrapair paternities in black-browed *Thalassarche melanophris*, grey-headed *T. chrysostoma* and wandering albatrosses *Diomedea exulans* at South Georgia. *Journal of Avian Biology*, 37, 331–338.

Burgess, M. D., Smith, K. W., Evans, K. L., Leech, D., Pearce-Higgins, J. W., et al (2018). Tritrophic phenological match–mismatch in space and time. *Nature Ecology & Evolution*, 2(6), 970–975.

Butchart, S. H. M., Seddon, N. & Ekstrom, J. M. M. (1999). Polyandry and competition for territories in bronze-winged jacanas. *Journal of Animal Ecology*, 68, 928–939.

Caspers, B. A., Hagelin, J. C., Paul, M., Bock, S., Willeke, S. & Krause, E. T. (2017). Zebra finch chicks recognise parental scent, and retain chemosensory knowledge of their genetic mother, even after egg cross-fostering. *Scientific Reports*, 7, 12859.

Colbourne, R. (2002). Incubation behaviour and egg physiology of kiwi (*Apteryx spp.*) in natural habitats. *New Zealand Journal of Ecology*, 26, 129–138.

Covas, R. (2012). Evolution of reproductive life histories in island birds worldwide. *Proceedings of the Royal Society B*, 279, 1531–1537.

Cuervo, J. J., Luisa, M. & Pape Mbller, A. (1999). Phenotypic variation and fluctuating asymmetry in sexually dimorphic feather ornaments in relation to sex and mating system. *Biological Journal of the Linnean Society* (Vol. 68). 505–529.

Cusick, J. A., de Villa, M., DuVal, E. H. & Cox, J. A. (2018). How do helpers help? Helper contributions throughout the nesting cycle in the cooperatively breeding brown-headed nuthatch. *Behavioral Ecology and Sociobiology*, 72(3), 43.

Diamond, J. & Bond, A. B. (2010). Social Behavior and the Ontogeny of Foraging in the Kea (*Nestor notabilis*). *Ethology*, 88(2), 128–144.

Drummond, H., Ramos, A. G., S Anchez-Macouzet, O. & Rodríguez, C. (2016). An unsuspected cost of mate familiarity: increased loss of paternity. *Animal Behaviour*, 111, 213–216.

Duval, E. H., Vanderbilt, C. C. & M'gonigle, L. K. (2018). The spatial dynamics of female choice in an exploded lek generate benefits of aggregation for experienced males. *Animal Behaviour*, 143, 215–225.

Emlen, S. T. & Oring, L. W. (1977). Ecology, sexual selection, and the evolution of mating systems. *Science*, 197, 215–223.

Emlen, S. T. & Wrege, P. H. (2004). Size dimorphism, intrasexual competition, and sexual selection in wattled jacana (*Jacana jacana*), a sex-role-reversed shorebird in Panama. *The Auk*, 121(2), 391–403.

Emlen, S. T., Wrege, P. H. & Webster, M. S. (1998). Cuckoldry as a cost of polyandry in the sex-role-reversed wattled jacana, *Jacana jacana*. *Proceedings of the Royal Society B*, 265, 2359–2364.

English, P. A. & Montgomerie, R. (2011). Robin's egg blue: does egg color influence male parental care? *Behavioral Ecology and Sociobiology*, 65(5), 1029–1036.

Evans, S. R. & Gustafsson, L. (2017). Climate change upends selection on ornamentation in a wild bird. *Nature Ecology & Evolution*, 1, 39.

Fargallo, J., De León, A. & Potti, J. (2001). Nest-maintenance effort and health status in chinstrap penguins, *Pygoscelis antarctica*: the functional significance of stone-provisioning behaviour. *Behavioral Ecology and Sociobiology*, 50(2), 141–150.

Feeney, W. E., Welbergen, J. A. & Langmore, N. E. (2012). The frontline of avian brood parasite-host coevolution. *Animal Behaviour*.

Foerster, K., Delhey, K., Johnsen, A., Lifjeld, J. & Kempenaers, B. (2003). Females increase offspring heterozygosity and fitness through extra-pair matings. *Nature*, 425, 714–717.

Fromhage, L. & Jennions, M. D. (2016). Coevolution of parental investment and sexually selected traits drives sex-role divergence. *Nature Communications*, 7, 1251.

Fujiwara, H. E., Kanesada, A., Okamoto, Y., Satoh, R., Watanabe, A. & Miyamoto, T. (2011). Long-term maintenance and eventual extinction of preference for a mate's call in the female budgerigar. *Animal Behaviour*, 82, 971–979.

Geberzahn, N., Goymann, W., Muck, C. & Ten Cate, C. (2009). Females alter their song when challenged in a sex-role reversed bird species. *Behavioral Ecology and Sociobiology*, 64, 193– 204.

Geberzahn, N., Goymann, W. & Ten Cate, C. (2010). Threat signaling in female song-evidence from playbacks in a sex-role reversed bird species. *Behavioral Ecology*, *21*, 1147–1155.

Goth, A. (2007). Mound and mate choice in a polyandrous megapode: females lay more and larger eggs in nesting mounds with the best incubation temperatures. *The Auk*, *124*, 253–263.

Goymann, W. & Wingfield, J. C. (2004). Competing females and caring males. Sex steroids in African black coucals, *Centropus grillii*. *Animal Behaviour*, *68*, 733–740.

Green, J. P., Freckleton, R. P. & Hatchwell, B. J. (2016). Variation in helper effort among cooperatively breeding bird species is consistent with Hamilton's Rule. *Nature Communications*, *7*(1), 12663.

Griesser, Michael, Drobniak, S. M., Nakagawa, S. & Botero, C. A. (2017). Family living sets the stage for cooperative breeding and ecological resilience in birds. *PLOS Biology*, *15*(6), e2000483.

Hagelin, J. C., Jones, I. L. & Rasmussen, L. E. L. (2003). A tangerine-scented social odour in a monogamous seabird. *Proceedings of the Royal Society B*, *270*, 1323–1329.

Hajduk, G. K., Cockburn, A., Margraf, N., Osmond, H. L., Walling, C. A. & Kruuk, L. E. B. (2018). Inbreeding, inbreeding depression, and infidelity in a cooperatively breeding bird. *Evolution*, *72*, 1500–1514.

Harts, A. M. F., Booksmythe, I. & Jennions, M. D. (2016). Mate guarding and frequent copulation in birds: A meta-analysis of their relationship to paternity and male phenotype. *Evolution*, *70*(12), 2789–2808.

Hasselquist, D. (2001). Social mating systems and extrapair fertilizations in passerine birds. *Behavioral Ecology*, *12*, 457–466.

Hawk, G., Delay, L. S., Faaborg, J., Naranjo, J., Paz, S. M., De Vries, T. & Parker, P. G. (1996). Paternal Care in the Cooperatively Polyandrous. *The Condor*, *98*.

Hayes, M. A., Britten, H. B. & Barzen, J. A. (2006). Extra-pair fertilizations in sandhill cranes revealed using microsatellite DNA markers. *The Condor*, *108*, 970–976.

Heg, D. & Van Treuren, R. (1998). Female-female cooperation in polygynous oystercatchers. *Nature*, *319*, 687–691.

Heg, Dik, Bruinzeel, L. W. & Ens, B. J. (2003). Fitness consequences of divorce in the oystercatcher, *Haematopus ostralegus*. *Animal Behaviour*, *66*, 175–184.

Heinsohn, R., Ebert, D., Legge, S. & Peakall, R. (2007). Genetic evidence for cooperative polyandry in reverse dichromatic Eclectus parrots. *Animal Behaviour*, *74*, 1047–1054.

Horseman, N. D. & Buntin, J. D. (1995). Regulation of Pigeon Cropmilk Secretion and Parental Behaviors by Prolactin. *Annual Review of Nutrition*, *15*(1), 213–238.

Ihle, M., Kempenaers, B. & Forstmeier, W. (2015). Fitness Benefits of Mate Choice for Compatibility in a Socially Monogamous Species. *PLOS Biology*, *13*, e1002248.

Ismar, S. M. H., Daniel, C., Stephenson, B. M. & Hauber, M. E. (2010). Mate replacement entails a fitness cost for a socially monogamous seabird. *Naturwissenschaften*, *97*, 109– 113.

Johnsen, T. S. & Zuk, M. (1996). Repeatability of mate choice in female red jungle fowl. *Behavioral Ecology* (Vol. 7).

Jose, J., Soler, À., Pape Méller, A. & Soler, M. (1998). Nest building, sexual selection and parental investment. *Evolutionary Ecology*, *12*, 427–444.

Jouventin, P., Charmantier, A., Dubois, M.-P., Jarne, P. & Bried, J. (2006). Extra-pair paternity in the strongly monogamous Wandering Albatross *Diomedea exulans* has no apparent benefits for females. *Ibis*, *149*(1), 67–78.

Kenny, E., Birkhead, T. R. & Green, J. P. (2017). Allopreening in birds. *Behavioral Ecology*, *28*(4), 1142–1148.

Kilner, R. M. (2006). The evolution of egg colour and patterning in birds. *Biological Reviews*, *81*(03), 383–406.

Kokko, H., Gunnarsson, T. G., Morrell, L. J. & Gill, J. A. (2006). Why do female migratory birds arrive later than males? *Journal of Animal Ecology*, *75*, 1293–1303.

Kraaijeveld, K., Gregurke, J., Hall, C., Komdeur, J. & Mulder, R. A. (2004). Mutual ornamentation, sexual selection, and social dominance in the black swan. *Behavioral Ecology*, *15*, 380–389.

Lank, D. B., Smith, C. M., Hanotte, O., Ohtonen, A., Bailey, S. & Burke, T. (2002). High frequency of polyandry in a lek mating system. *Behavioral Ecology*, *13*(2), 209–215.

Leader, N. & Yom-Tov, Y. (1998). The possible function of stone ramparts at the nest entrance of the blackstart. *Animal Behaviour*, *56*(1), 207–217.

Ledwon, M. & Neubauer, G. (2018). True deception during extra-pair courtship feeding: cheating whiskered tern *Chlidonias hybrida* females perform better. *Journal of Avian Biology*, *49*(6).

Leighton, G. M., Echeverri, S., Heinrich, D., & Kolberg, H. (2015). Relatedness predicts multiple measures of investment in cooperative nest construction in sociable weavers. *Behavioral Ecology and Sociobiology*, *69*, 1835–1843.

Leniowski, K. & Wegrzyn, E. (2018). Synchronisation of parental behaviours reduces the risk of nest predation in a socially monogamous passerine bird. *Scientific Reports*, *8*, 7385.

Levey, D. J., Duncan, R. S. & Levins, C. F. (2004). Use of dung as a tool by burrowing owls. *Nature*, *431*(7004), 39–39.

Liker, A., Freckleton, R. P. & Székely, T. (2013). The evolution of sex roles in birds is related to adult sex ratio. *Nature Communications*, *4*.

Limmer, B. & Becker, P. H. (2010). Improvement of reproductive performance with age and breeding experience depends on recruitment age in a long-lived seabird. *Oikos*, *119*(3), 500–507.

Lovell, P. G., Ruxton, G. D., Langridge, K. V & Spencer, K. A. (2013). Egg-laying substrate selection for optimal camouflage by quail. *Current Biology*, *23*(3), 260–264.

Lyon, B. E. & Shizuka, D. (2010). Communal Breeding: Clever Defense Against Cheats. *Current Biology*, *20*, R931–R933.

Magrath, M. J. L. & Komdeur, J. (2003). Is male care compromised by additional mating opportunity? *Trends in Ecology and Evolution*, *18*, 424–430.

Maness, T. J. & Anderson, D. J. (2008). Mate rotation by female choice and coercive divorce in Nazca boobies, *Sula granti*. *Animal Behaviour*, *76*, 1267–1277.

Mariette, M. M. & Griffith, S. C. (2015). The Adaptive Significance of Provisioning and Foraging Coordination between Breeding Partners. *American Naturalist*, *185*(2).

Martín-Vivaldi, M., Soler, J. J., Peralta-Sánchez, J. M., Arco, L., Martín-Platero, A. M., et al. (2014). Special structures of hoopoe eggshells enhance the adhesion of symbiont-carrying uropygial secretion that increase hatching success. *Journal of Animal Ecology*, *83*(6), 1289–1301.

Martin, T. E., Boyce, A. J., Fierro-Calderón, K., Mitchell, A. E., Armstad, C. E., Mouton, J. C. & Bin Soudi, E. E. (2017). Enclosed nests may provide greater thermal than nest predation benefits compared with open nests across latitudes. *Functional Ecology*, *31*(6), 1231–1240.

Martin, T. E., Scott, J. & Menge, C. (2000). Nest predation increases with parental activity: separating nest site and parental activity effects. *Proceedings of the Royal Society B*, *267*, 2287–2293.

Massaro, M., Sainudiin, R., Merton, D., Briskie, J. V & Poole, A. M. (2013). Human-Assisted Spread of a Maladaptive Behavior in a Critically Endangered Bird. *PLoS ONE*, *8*(12), 79066.

Massaro, M., Starling-Windhof, A., Briskie, J. V & Martin, T. E. (2008). Introduced mammalian predators induce behavioural changes in parental care in an endemic New Zealand bird. *PLoS ONE, 3*(6), 2331.

Massen, J. J. M., Szipl, G., Spreafico, M., & Bugnyar, T. (2014). Ravens intervene in others' bonding attempts. *Current Biology, 24,* 2733–2736.

Matysiokova, B. & Remeš, V. (2013). Faithful females receive more help: the extent of male parental care during incubation in relation to extra-pair paternity in songbirds. *Journal of Evolutionary Biology, 26,* 155–162.

Maurer, G., Portugal, S. J., Hauber, M. E., Mikšík, I., Russell, D. G. D. & Cassey, P. (2015). First light for avian embryos: eggshell thickness and pigmentation mediate variation in development and UV exposure in wild bird eggs. *Functional Ecology, 29*(2), 209–218.

Medina, I. & Langmore, N. E. (2016). The evolution of acceptance and tolerance in hosts of avian brood parasites. *Biological Reviews, 91,* 569–577.

Miles, M. C. & Fuxjager, M. J. (2018). Synergistic selection regimens drive the evolution of display complexity in birds of paradise. *Journal of Animal Ecology, 87*(4), 1149–1159.

Møller, A. P., Kimball, · R T & Erritzøe, · J. (1996). Sexual ornamentation, condition, and immune defence in the house sparrow *Passer domesticus. Behavioral Ecology and Sociobiology, 39,* 317–322.

Moravec, M. L., Striedter, G. F. & Burley, N. T. (2006). Assortative Pairing Based on Contact Call Similarity in Budgerigars, *Melopsittacus undulatus. Ethology, 112*(11), 1108–1116.

Mougeot, F., Thibault, J.-C. & Bretagnolle, V. (2002). Effects of territorial intrusions, courtship feedings and mate fidelity on the copulation behaviour of the osprey. *Animal Behaviour, 64,* 759–769.

Muck, C. & Goymann, W. (2018). Exogenous testosterone does not modulate aggression in sex-role-reversed female Barred Buttonquails, *Turnix suscitator. Journal of Ornithology,* 1–9.

Muck, C., Kempenaers, B., Kuhn, S., Valcu, M. & Goymann, W. (2009). Paternity in the classical polyandrous black coucal (*Centropus grillii*)—a cuckoo accepting cuckoldry? *Behavioral Ecology, 20,* 1185–1193.

Murphy, T. G., Hernández-Muciño, D., Osorio-Beristain, M., Montgomerie, R. & Omland, K. E. (2009). Carotenoid-based status signaling by females in the tropical streak-backed oriole. *Behavioral Ecology, 20,* 1000–1006.

Olson, V. A. & Turvey, S. T. (2013). The evolution of sexual dimorphism in New Zealand giant moa (*Dinornis*) and other ratites. *Proceedings of the Royal Society B: Biological Sciences.*

Ouyang, J. Q., Van Oers, K., Quetting, M., & Hau, M. (2014). Becoming more like your mate: hormonal similarity reduces divorce rates in a wild songbird. *Animal Behaviour, 98,* 87–93.

Owens, I. P. F. (2002). Male-only care and classical polyandry in birds: phylogeny, ecology and sex differences in remating opportunities. *Philosophical Transactions of the Royal Society of London. Series B, Biological Sciences, 357*(1419), 283–293.

Owens, I. P. F., Dixon, A., Burke, T. & Thompson, D. B. A. (1995). Strategic paternity assurance in the sex-role reversed Eurasian dotterel (*Charadrius morinellus*): behavioral and genetic evidence. *Behavioral Ecology, 6,* 14–21.

Pechacek, P., Michalek, K. G., Winkler, H., & Blomqvist, D. (2006). Classical polyandry found in the three-toed woodpecker *Picoides tridactylus. Journal of Ornithology, 147,* 112–114.

Phillips, J. N. & Derryberry, E. P. (2017). Vocal performance is a salient signal for male-male competition in White-crowned Sparrows. *The Auk, 134,* 564–574.

Pizzari, T., Cornwallis, C. K., Lovlie, H., et al. (2003). Sophisticated sperm allocation in male fowl. *Nature, 426,* 70–74.

Pogány, Á., Szentirmai, I., Komdeur, J. & Székely, T. (2008). Sexual conflict and consistency of offspring desertion in Eurasian penduline tit *Remiz pendulinus. BMC Evolutionary Biology, 8*(1), 242.

Powlesland, R. G., Lloyd, B. D., Best, H. A. & Merton, D. V. (2008). Breeding biology of the Kakapo *Strigops habroptilus* on Stewart Island, New Zealand. *Ibis, 134*(4), 361–373.

Price, J. J. (2009). Evolution and life-history correlates of female song in the New World blackbirds. *Behavioral Ecology, 20,* 967–977.

Pruett-Jones, S. (1992). Independent Versus Nonindependent Mate Choice: Do Females Copy Each Other? *The American Naturalist* (Vol. 140).

Quader, S. (2006). What makes a good nest? Benefits of nest choice to female baya weavers (*Ploceus philippinus*). *The Auk, 123,* 475–486.

Queller, D. C. (1997). Why do Females Care More than Males? *Proceedings of the Royal Society B, 264*(1388), 1555–1557.

Quinn, J. S., Haselmayer, J., Dey, C. & Jamieson, I. G. (2012). Tolerance of female co-breeders in joint-laying pukeko: the role of egg recognition and peace incentives. *Animal Behaviour, 83,* 1035–1041.

Rico-Guevara, A. & Araya-Salas, M. (2015). Bills as daggers? A test for sexually dimorphic weapons in a lekking hummingbird. *Behavioral Ecology, 26,* 21–22.

Riehl, C. (2010). A simple rule reduces costs of extragroup parasitism in a communally breeding bird. *Current Biology, 20,* 1830–1833.

Riehl, C. (2011). Living with strangers: direct benefits favour non-kin cooperation in a communally nesting bird. *Proceedings of the Royal Society B, 278,* 1728–1735.

Riehl, C. (2012). Mating system and reproductive skew in a communally breeding cuckoo: hard-working males do not sire more young. *Animal Behaviour, 84,* 707–714.

Riehl, C. (2013). Evolutionary routes to non-kin cooperative breeding in birds. *Proceedings of the Royal Society B, 280,* 0132245.

Riehl, C. (2017). Kinship and Incest Avoidance Drive Patterns of Reproductive Skew in Cooperatively Breeding Birds. *American Naturalist, 190,* 774–785.

Riehl, C. & Strong, M. J. (2018). Stable social relationships between unrelated females increase individual fitness in a cooperative bird. *Proceedings of the Royal Society B, 285,* 20180130.

Riehl, C. & Strong, M. J. (2019). Social parasitism as an alternative reproductive tactic in a cooperatively breeding cuckoo. *Nature, 567,* 96–99.

Riehl, C., Strong, M. J. & Edwards, S. V. (2015). Inferential reasoning and egg rejection in a cooperatively breeding cuckoo. *Animal Cognition, 18,* 75–82.

Robertson, J. K., Caldwell, J. R., Grieves, L. A., Samuelsen, A., Schmaltz, G. S. & Quinn, J. S. (2018). Male parental effort predicts reproductive contribution in the joint-nesting, Smooth-billed Ani (*Crotophaga ani*). *Journal of Ornithology, 159,* 471–481.

Rossi, M., Marfull, R., Golüke, S., Komdeur, J., Korsten, P. & Caspers, B. A. (2017). Begging blue tit nestlings discriminate between the odour of familiar and unfamiliar conspecifics. *Functional Ecology, 31*(9), 1761–1769.

Rowe, M., Bakst, M. R. & Pruett-Jones, S. (2008). Good vibrations? Structure and function of the cloacal tip of male Australian Maluridae. *Journal of Avian Biology, 39*(3), 348–354.

Rytknen, S., Orell, M. & Koivula, K. (1993). Sex-role reversal in willow tit nest defence. *Behavioral Ecology and Sociobiology, 33*, 275–282.

Sánchez-Lafuente, A. M., Alcántara, J. M. & Romero, M. (1998). Nest-Site Selection and Nest Predation in the Purple Swamphen. *Journal of Field Ornithology, 69*, 563–576.

Schamel, D., Tracy, D. M., Lank, D. B. & Westneat, D. F. (2004). Mate guarding, copulation strategies and paternity in the sex-role reversed, socially polyandrous red-necked phalarope *Phalaropus lobatus*. *Behavioral Ecology and Sociobiology, 57*, 110–118.

Schmaltz, G., Quinn, J. S. & Lentz, C. (2008). Competition and waste in the communally breeding smooth-billed ani: effects of group size on egg-laying behaviour. *Animal Behaviour, 76*, 153–162.

Schuetz, J. G. (2005). Common waxbills use carnivore scat to reduce the risk of nest predation. *Behavioral Ecology, 15*, 133–137.

Shen, S.-F., Emlen, S. T., Koenig, W. D. & Rubenstein, D. R. (2017). The ecology of cooperative breeding behaviour. *Ecology Letters, 20*(6), 708–720.

Slatyer, R. A., Mautz, B. S., Backwell, P. R. Y. & Jennions, M. D. (2012). Estimating genetic benefits of polyandry from experimental studies: A meta-analysis. *Biological Reviews, 87*, 1–33.

Soler, J. J. & Soler, M. (2017). Evolutionary change: facultative virulence by brood parasites and tolerance and plastic resistance by hosts. *Animal Behaviour, 125*, 101–107.

Soler, M. & de Neve, L. (2013). Brood mate eviction or brood mate acceptance by brood parasitic nestlings? An experimental study with the non-evictor great spotted cuckoo and its magpie host. *Behavioral Ecology and Sociobiology, 67*, 601– 607.

Soler, M., de Neve, L., Roldán, M., Macías-Sánchez, E. & Martín-Gálvez, D. (2012). Do great spotted cuckoo nestlings beg dishonestly? *Animal Behaviour, 83*, 163–169.

Sorenson, M. D. & Payne, R. B. (2002). Molecular Genetic Perspectives on Avian Brood Parasitism. *Integrative and Comparative Biology, 42*, 388–400.

Sorenson, M. D. & Payne, R. B. (2001). A single ancient origin of brood parasitism in African finches: Implications for host-parasite coevolution. *Evolution, 5*, 2550–2567.

Sorenson, M. D., Sefc, K. M. & Payne, R. B. (2003). Speciation by host switch in brood parasitic indigobirds. *Nature, 424*, 928–931.

Spoon, T. R., Millam, J. R. & Owings, D. H. (2006). The importance of mate behavioural compatibility in parenting and reproductive success by cockatiels, *Nymphicus hollandicus*. *Animal Behaviour, 71*, 315-326.

Spottiswoode, Claire N., Stryjewski, K. F., Quader, S., Colebrook-Robjent, J. F. R. & Sorenson, M. D. (2011). Ancient host specificity within a single species of brood parasitic bird. *Proceedings of the National Academy of Sciences, 108*, 17738–17742.

Spottiswoode, Claire N. (2013). A brood parasite selects for its own egg traits. *Biology Letters, 9*, 20130573.

Spottiswoode, Claire N. & Koorevaar, J. (2012). A stab in the dark: Chick killing by brood parasitic honeyguides. *Biology Letters, 8*, 241–244.

Spottiswoode, Claire N. & Stevens, M. (2011). How to evade a coevolving brood parasite: Egg discrimination versus egg variability as host defences. *Proceedings of the Royal Society B: Biological Sciences, 278*, 3566–3573.

Spottiswoode, Claire N. & Stevens, M. (2012). Host-Parasite Arms Races and Rapid Changes in Bird Egg Appearance. *The American Naturalist, 179*, 633–648.

Stoddard, M. C. & Hauber, M. E. (2017). Colour, vision and coevolution in avian brood parasitism. *Philosophical Transactions of the Royal Society B: Biological Sciences, 372*, 20160339.

Teitelbaum, C. S., Converse, S. J. & Mueller, T. (2017). Birds choose long-term partners years before breeding. *Animal Behaviour, 134*, 147–154.

Theuerkauf, J., Kuehn, R., Rouys, S., Bloc, H. & Gula, R. (2018). Fraternal Polyandry and Clannish Spatial Organization in a Flightless Island Bird. *Current Biology, 28*, 1482–1488.

Thorogood, R. & Davies, N. B. (2012). Cuckoos combat socially transmitted defenses of reed warbler hosts with a plumage polymorphism. *Science, 337*, 578–580.

Van Dijk, R. E., Székely, T. S., Komdeur, J., Pogány, A., Fawcett, T. W. & Weissing, F. J. (2012). Individual variation and the resolution of conflict over parental care in penduline tits. *Proceedings of the Royal Society B, 279*, 1927–1936.

Vehrencamp, S. L. (2000). Evolutionary routes to joint-female nesting in birds. *Behavioral Ecology, 11*(3), 334–344.

Voigt, C. (2016). Neuroendocrine correlates of sex-role reversal in barred buttonquails. *Proceedings of the Royal Society B, 283*, 20161969.

Warning, N. & Benedict, L. (2015). Paving the way: Multifunctional nest architecture of the Rock Wren. *The Auk, 132*(1), 288–299.

Welbergen, J. A. & Davies, N. B. (2011). A parasite in wolf's clothing: Hawk mimicry reduces mobbing of cuckoos by hosts. *Behavioral Ecology, 22*, 574–579.

Welbergen, J. A. & Davies, N. B. (2012). Direct and indirect assessment of parasitism risk by a cuckoo host. *Behavioral Ecology, 23*, 783–789.

Wiebe, K. L. & Kempenaers, B. (2009). The social and genetic mating system in flickers linked to partially reversed sex roles. *Behavioral Ecology, 20*, 453–458.

Wiemann, J., Yang, T.-R., Sander, P. N., Schneider, M., Engeser, M., et al (2017). Dinosaur origin of egg color: oviraptors laid blue-green eggs. *PeerJ—Life and Environmental Sciences, 5*, e3706.

Young, L. C. & Vanderwerf, E. A. (2014). Adaptive value of same-sex pairing in Laysan albatross. *Proceedings of the Royal Society B, 281*, 20132473.

Zanette, L. Y., White, A. F., Allen, M. C. & Clinchy, M. (2011). Perceived predation risk reduces the number of offspring songbirds produce per year. *Science, 334*, 1398–1401.

Zheng, J., Li, D. & Zhang, Z. (2018). Breeding biology and parental care strategy of the little-known Chinese Penduline Tit (*Remiz consobrinus*). *Journal of Ornithology, 159*(3), 657–666.

INDEX

ACKNOWLEDGMENTS

Books, like birds, have long been my passion and inspiration, allowing me to enter communities that transcend physical boundaries of geography or biology.

I owe more than I can ever express to the organisms (including humans) that have contributed to this book. From Ivy Press, I am indebted to Kate Shanahan for enthusiastically accepting my idea of writing about the light and dark sides of bird family life. Also to Tom Kitch, Wayne Blades, James Lawrence, and the entire art team that produce the visually stunning books Ivy Press is famous for. Jo Bentley has been an enormously patient and insightful editor. I am also grateful to Liz Drewitt, Mike Webster, Nancy Curtis, Aileen Nielsen, CJ Huang, and an expert reviewer for PUP for reading the book in its entirety, and for improving it.

This book is about family life, and I have an extended family to thank. My friends and mentors from the Nature Society (Singapore), EEB at Princeton, OEB and the MCZ at Harvard, Zoology and Darwin College at Cambridge, OBE and Forestry at the University of Montana, the MNHC, Nature Kenya, Mpala Research Centre, and colleagues from Biology at the University of Alaska have all been immeasurably supportive. In particular, David Haig and the FIAT group are a source of wise insights and ideas, many of which inspired this book. Doug Emlen put me in touch with Ivy Press in the first place, and has been a true mentor in both science and science communication. I am also deeply grateful to friends, students, and guests, from both Cambridges, Kenya, Zambia, Montana, and Anchorage, on whom I have tried out so many of the stories in this book.

Most of all, I thank my parents, my husband Jesse Weber, and Ana, for so patiently putting up with my love for birds, and for whom this book is written.

PICTURE CREDITS

The publisher would like to thank the following for permission to reproduce copyright material:

Alamy Stock Photo/Agami Photo Agency 64, 126r, 185t, Andrew Walmsley 156, Arco Images GmbH 112, 161r, Auscape International Pty Ltd 86b, Avalon/Photoshot License 46b, Bill Coster 92, BIOSPHOTO 133t, blickwinkel 69b, BSIP SA 84b, Buiten-Beeld 130, Daniel Borzynski 58l, David Tipling Photo Library 162, 175t, David Whitaker 51b, Dirk Reuter 179b, Drew Buckley 182br, Duncan Usher 29t, Florian Schulz 127, FLPA 55, 56tl, franzfoto.com 42tl, Genevieve Vallee 148, Hal Beral/VWPics 104t, Hans Lang/ImageBROKER 113, ImageBROKER 65, 177b, John Trevor Platt 146, Juanma Pelegrin 81b, Karen Debler 33b, Konrad Wothe/Minden Pictures 41, Leonardo Mercon/VWPics 161l, Lewis Thomson 172, Martin Smart 60, Mauritius Images GmbH 68c, Minden Pictures 49t, 51t, 131, 137t, 153, 165t, 165b, 179t, 184, Nature Photographers Ltd 173tc (eggs), 173tr (eggs), 173bl (eggs), Nature Picture Library 21b, 72, Neal Mishler 117, Oyvind Martinsen-Panama Wildlife 42cr, Remo Savisaar 182bl, Rick & Nora Bowers 121b, Robert Harding 138, Rolf Nussbaumer 147t, Steve Bloom Images 23, Steven Blandin 123t, The Natural History Museum 120, Tobias Peciva 101b, Wildlife GmbH 123b, William Leaman 140. Andy Reago & Chrissy McClarren 58r. Ardea.com/Jean-Paul Ferrero 136, Jim Zipp 22b, M Watson 56tr, Peter Steyn 175br, Stefan Meyers 31. Australian Antarctic Division/Christopher Wilson 93. Bernard Spragg 30b. Beverley Van Praagh 45tr. Charlie Davies 145, 154. Claire Spottiswoode 169t, 177t. F Sergio 79. Getty Images/Alexandra Rudge 157, Auscape/Universal Images Group 94, Bernd Zoller 102t, Chilkoot 18-19, Diana Robinson Photography 111, drferry 53t, Education Images 40t, Fabian von Poser 86t, Foto4440 36-7, GlobalP 57b, HankDuh 160t, Harry Eggens 63, James Hager 144b, Jean Raymond Gammino/EyeEm 81t, Jim Cumming 115, Mary Plage 116tl, Sandergroffen 73t, sherjaca 57t Thomas D Mcavoy/The LIFE Picture Collection 98t, Vicki Jauron, Babylon and Beyond Photography 141, Westend61 80, 97b, Wild Horizons/Universal Images Group 98b, Will Gray 33t, Zahoor Salmi 135. Greg Schneider 107. High Arctic Institute/Kurt Burnham 76t. Ivy Press/Andrew Perris 11l. Justin Schuetz 178t. Kim Wormald 155. M C Stoddard/© NHM 173br (eggs), 173bc (eggs). Max Planck Gesellschaft/Wolfgang Goymann 128l. Naomi Langmore 178br. Nature Picture Library/Brent Stephenson 87, 110, Pete Oxford 150b, Roger Powell 42cl, Rolf Nussbaumer 42b. Paul Krawczuk 32b. Pixabay/Per Nicolaisen 125, Primalimage 45b, Sid Litke 83i. Rahul & Khushboo Sharma 45tl. Roger Wasley 19l. rspb-images.com/Danny Green 142b, David Kjaer 77, David Osborn 59, Mike Lane 62, 78t, Ray Kennedy 53b, 158t, Roger Tidman 164. Ryan Poplin 38. Sandra Pond 48-9b. Shutterstock/Agami Photo Agency 149, Amelia Martin 168b, Andrew M Allport 100, ArCaLu 116tr, Aria R J Warren 70, bearacreative 173tr (bird), Beate Wolter 134, Bildagentur Zoonar GmbH 91, Bird Butterfly Wildlife 9, BlueOrange Studio 16, Bob Hilscher 89, Bob Pool 118, Borislav Borisov 7, Brian Lasenby back cover, 103, Butterfly Hunter 132, Cat Sparks 156t, Cathy Withers-Clarke 73t, Cezary Korkosz 14, Ciobaniuc Adrian Eugen 147b, clarst5 24, 173tcr (bird), Clea Calderoni 46t, Coulanges 106, David McMillan 83, David Steele 158b, Dawie Nolte 175bl, Dennis Jacobsen 40b, Dolores Giraldez Alonso 99l, Don Mammoser 22t, Eric Isselee 11r, 166b, 168t, Erni 26, Evgeni Stefanov 2, fivespots 104b, FloridaStock 15, 76b, Foto 4440 170, FotoRequest 25, Francois Loubser 44, Gelpi 143, Georgios Alexandris 173br (bird), Giedriius 109t, godi photo 84t, Greg & Jan Ritchie 8, Ian Maton 121t, iliuta goean 176l, Jaagurak 82, JaklZdenek 61, Jenni Maija Helena 85, Jody Ann 167, John L Absher 102b, Jukka Jantunen 54, Keneva Photography 109b, Kiki Dohmeier 19r, Laura J P Richardson 34-4, LouieLea 160b, Luiz Antonio da Silva 88b, Luke Shelley 48, Mark Medcalf 47, Mark Robert Paton 176r, Martin Pelanek 42tr, 126l, Menno Schaefer 50, meunierd 152, Michael W NZ 137b, milan saju k 74t, mohd siberi 174, Ondrej Prosicky 68t, Pablo Rodriguez Merkel 124t, 129, 180t, Paul Wittet 116b, Peter Bocklandt 101t, Petr Simon 90, phototr 78b, Popova Valeriya 56b, Rachata Kietsirikul 150t, Raul Topan 166t, Reimar 97t, Ryzhkov Sergey 30t, Shaun Jeffers front cover, Simon_g 171, Sompreaw 27, Stacey Ann Alberts 151, Steve Byland 185b, Syed F Abbas 69t, tahirsphotography 66, Tim Zurowski 88t, 144t, Torii Lynn Weaver 10, Torsten Pursche 128r, Uwe Bergwitz 71, Vincenzo Iacovoni 124b, Vishnevskiy Vasily 52, 173bl (egg), Vitaly Ilyasoz 173bl (bird), Wang LiQiang 43, 181t, 182t, Wildlife World 74b, witezz77 173bc (bird), Wright Out There 142t, Wulong Tommy 96. Sushyue Liao 29b. USFWS/Pacific Region 99r, Tom Koerner 133b. Vittorio Baglione 181b. Wenfei Tong 12, 20, 21t, 75, 169. Wikimedia/Collection of Jacques Perrin de Brichambaut/Museum de Toulouse 173l (egg r), New Jersey Birds 183, Sharp Photography 180b, Steve Berardi 68b. Wolfgang Forstmeier 32t.

All reasonable efforts have been made to trace copyright holders and to obtain their permission for the use of copyright material. The publisher apologizes for any errors or omissions in the list above and will gratefully incorporate any corrections in future reprints if notified.

Figure acknowledgments:

p13 after Brusatte, S. L., O'Connor, J. K. & Jarvis, E. D., The Origin and Diversification of Birds, *Current Biology* 25, Issue 19, October 2015, pp. R888-R898; p105 after a concept by David Haig; p140 & 171 after Feeney, W. et al, Brood Parasitism and the Evolution of Cooperative Breeding in Birds, *Science* 342 (2013), pp. 1506-8.